Introduction to Forestry Economics

Peter H. Pearse

University of British Columbia Press
Vancouver 1990

ISBN 0-7748-0336-3

∞

Printed in Canada on acid-free paper

This book has been published
with the help of a grant from
the Department of Communications.

Canadian Cataloguing in Publication Data

Pearse, Peter H., 1932–
Introduction to forestry economics

Includes bibliographical references.
ISBN 0-7748-0336-3

1. Forests and forestry – Economic aspects.
I. Title.
SD393.P42 1990 338.1'7498 C89-091617-9

Dedicated to

Anthony Scott
mentor, colleague, and friend

Contents

Tables and Figures

TABLES

FIGURES

Foreword

Professor Pearse has written a much-needed textbook for introductory forestry economists. There are other textbooks and an extensive technical literature on forestry economics, but, in my own teaching experience, I have found that they meet the needs neither of first- and second-year undergraduate forestry students nor of economics students interested in forest management and policy. Similarly, there has never been a satisfactory general introduction to economic principles underlying most forest management and policy issues. This book satisfies all these needs.

Introduction to Forestry Economics links economic principles with both private and public forestry decision-making. The text adheres to a few basic economic principles, including opportunity cost, equimarginal conditions, and consumer sovereignty. Its broad topical emphases feature resource allocation over time and justifications for public sector market intervention. Resource allocation over time is the classic private sector forestry problem. It has received much attention from both professional foresters and economists, but the exposition in this book is one of the few at an introductory level.

The current importance of market intervention justifies its greater attention here than in previous forestry textbooks. Thus, there are chapters on valuing unpriced services, multiple use, property rights, and land tenure systems. Tenure is a significant topic in Canada and the United States, but it is of paramount importance for forestry and rural development in Asia and Africa. The material on this and other topics is always presented in a general manner meaningful for North American students and also useful for students from other backgrounds. Each chapter lists additional references for those who want more detailed information on particular topics.

Professor Pearse is well-qualified to write *Introduction to Forestry*

Economics. He has twice led Royal Commissions in Canada that examined forestry problems and issues, run for national elective office from an area where forests are of prime importance, and has been an adviser on natural resource development in several developing countries. Through these experiences he has earned a reputation as a clear expostulator of fundamental economics and forestry issues. Pearse also teaches economics and policy to undergraduate forestry students at the University of British Columbia and understands well the student demand for this book. I know no one with a better preparatory background for writing it.

William F. Hyde
Branch Chief, Economics Research Service, USDA,
and Associate Professor, Duke University

Preface

Forestry has attracted the interest of economists for more than two and a half centuries. Some of the great contributors to economic doctrine, notably the early German capital theorists, developed their concepts with reference to forestry. But we cannot claim the reverse; economic theory has not, historically, had much influence on forestry, at least not in North America.

To put it more strongly, economics has often been rejected by foresters. It has been seen as contrary to accepted forestry principles about how forests should be conserved and managed and how harvesting should be regulated.

True, courses in forest economics traditionally have had a place in the curricula of our professional forestry schools. But their place in the curriculum has typically been (to borrow a phrase) marginal. The subject has been treated by academic and professional foresters as one which must be acknowledged, but not permitted to impede good forestry policies and practices.

This textbook was motivated by a conviction that this view, that economics is not helpful to the practice of forestry, is mistaken. Forestry is, as I see it, the applied science of managing land and trees. We don't manage forests for their own sake, however; as far as they are concerned they are quite capable of managing themselves. Rather, we manage them to advance social objectives. We must have a purpose to which the effort is directed, and it may be industrial timber production, recreation, or a variety of other things of value to people. Economics is concerned with choices about how resources are allocated and used to create things of value to people. Having begun a career as a forester and later taken up economics, it seems to me that the two areas of study converge and complement one another. Forestry involves using land, labour, and capital to produce goods and services from forests, and economics helps in understanding how we can do this in ways that will best meet the needs of people.

Moreover, it is increasingly apparent that we cannot isolate forestry from the economic forces that drive other activities. The growing intensity and variety of demands on forests for recreation, aesthetic, and environmental benefits as well as timber give rise to complicated problems of choice among a wide range of human wants and needs - precisely the subject of economics. And as forestry in North America becomes more and more concerned with managing forests for various purposes rather than simply using them, it involves investments in direct competition with investment needs for other social purposes. So forestry must be understood in its full economic context, and this context can provide a unifying framework for analysing forestry problems.

Nevertheless, those of us who have taught courses in economics for undergraduate forestry students in the United States and Canada have often found it an uphill battle. Economic theories about the relative value of something today and something tomorrow, and how forestry investments must be compared with other investment opportunities, are received with scepticism. Courses in forest economics therefore must be designed to introduce the subject to reluctant students, and to convince them that it will be useful to them as forest managers.

For this purpose we need a textbook that helps students of forestry to understand the economic implications of the work they will do as professional foresters, bearing in mind that what they learn in one course on the subject will have to last most of them for many years. I have therefore made an effort, in writing this book, to distil the subject down to the fundamentals - the basic economic principles of forestry and how they bear on forest management and policy decisions.

Most textbooks available to undergraduate students of forestry economics seem to attempt too much. Some try to introduce the student to the principles of economics. But today a host of excellent introductory texts are available on this subject. Students of forestry economics have usually already taken a course in economic principles, and those who have not should be referred to suitable references. This book therefore begins from and builds on the general principles of economics, and elaborates on the particular concepts relevant to forest management.

At the same time an effort has been made to resist the temptation to elaborate on the basic theory, to cover all the qualifications and special cases, to explain the esoteric jargon that applies to them, and to digress on the taxes, regulations, and institutional circumstances of particular jurisdictions. Emphasis is on the concepts that all foresters should understand and remember throughout their professional careers. Those few who specialize in forestry economics, and become forest economists, will have ample opportunity to explore the ramifications of the subject in other courses and textbooks.

For the same reasons the use of mathematics, diagrams, and statistics has been reduced to the bare minimum. Students of forestry seem to grasp economic concepts more readily when they are presented logically and intuitively in the context of problems familiar to them.

I am content if my students completing the introductory course in forestry economics thoroughly understand a few fundamental concepts of economic theory and their relevance to forestry. Among these the principles of economic efficiency, opportunity cost, marginal analysis (which, incidentally, was developed with reference to forestry by early German theorists), and valuation over time are most important. Much of this book deals with the application of these concepts to problems of forestry.

Thus the preparation of this book has been, in large part, a process of winnowing through economic doctrine on one hand and forestry problems on the other in order to focus on the basic connections between the two. I have erred on the side of simplicity of theory and exposition, depending on instructors to guide their students toward relevant applications, practical problems, and further readings, some of which are suggested at the end of each chapter.

The book begins by gathering the threads of economics as they apply to forestry; the first two chapters sketch the scope of the subject and introduce the issues addressed in the rest of the book. The next three chapters introduce the variety of goods and services produced through forestry and how their values determine the most beneficial use of land and timber. Chapter 6 turns to the important principles of evaluation over time and techniques for assessing forest investments, which are applied to specific problems in subsequent chapters. These deal with such policy questions as the forest rotation age, the regulation of harvests over time, and property rights and taxes. The final chapter discusses some of the new analytical techniques for investigating issues of forest economics.

Both forestry and economics have historically been dominated by men. Belatedly, this situation has begun to change, and we should encourage the increasing involvement of women in these fields. In writing this book I have therefore been reluctant to use only the masculine in referring to the third person singular, and I have done so simply to avoid the cumbersome use of compound personal pronouns throughout the exposition.

Some of the manuscript for this book was written on an island in the Strait of Georgia, between British Columbia and the State of Washington. There, forests dominate the landscape, contributing importantly to aesthetic and recreational values as well as to employment in timber operations. The management of these forests raises most of the economic issues examined in this book. Sundry Island and Peavey Forest Products Limited, described in the prologues to the chapters, are only partly fictitious.

This book has benefited from the advice and criticism of many students and colleagues. I am particularly indebted to Dr. William Hyde of the U.S. Department of Agriculture and Duke University for comments on the manuscript, as well as for providing the foreword to the book. Professor Luis Constantino of the University of Alberta, also provided helpful advice and ideas, as did Professors Terry Heaps and David Haley, Ms Jeanette Lietch and Ms Cindy Pearce, and Mr. Michael Cragg. My colleague, Professor Anthony Scott, to whom this book is dedicated, provided continuing inspiration. The facilities of the Forest Economics and Policy Analysis Research Unit and the support of the Faculty of Forestry at the University of British Columbia made the whole project manageable. Preparation of the manuscript owes much to the expert assistance of Patsy Quay, Miriam Nachemia, Carmen Rida, and Sandra Buckingham. I must acknowledge, as well, the continuing guidance of my students over many years, who have honed my own appreciation of the applications of economics to forestry.

Peter H. Pearse
Vancouver, 1990

Introduction to Forestry Economics

CHAPTER ONE

Forestry's Economic Perspective

Out in the Forest . . .

Peavey Forest Products Limited owns a tract of timberland on Sundry Island, between Washington State and British Columbia. It is a relatively small company, producing sawlogs and pulpwood—mostly Douglas-fir, white fir, and red cedar. The company is run by its president, David Cameron, who makes all the final decisions about how much to harvest each month, what kind of logs to produce, how to organize production, and how much to spend on roadbuilding, equipment, and silviculture.

In making his decisions, Cameron's main objective is to generate profits for the company's shareholders. But his range of choice is limited. First, he can only produce what the forest is capable of producing. He depends on the company forester, Ian Olson, to advise him about the forestry and timber production possibilities on the company's lands. Second, he is constrained by laws and regulations about pollution control, worker safety, the use of public highways, fire protection, and a host of other things. These governmental restrictions are aimed at ensuring that his economic activities conform to the broader interests of society. Third, he has to produce the kind of logs that sawmills and pulpmills want to buy, and offer them for sale at prices no higher than the prices of other log producers in the region. This means that he must constantly strive to reduce costs and improve efficiency in order to successfully compete and generate maximum possible profits.

The company maintains a heavy investment in land, and in capital in the form of standing timber. Using these in combination with labour and other resources it produces timber products which other producers use as raw material to make final products like furniture and newspapers wanted by consumers. In this way, Peavey Forest Products Limited, like other enterprises in the economy, contributes to the material welfare of society as a whole.

Forestry calls on a variety of skills and disciplines of study. Professional foresters must combine knowledge drawn from biology and other natural sciences, applied sciences, and social sciences such as economics. Each discipline brings a different set of tools and methods to the task of managing forests. This book deals with forestry from an economic perspective.

This first chapter presents some context for studying forest economics. Those who must decide how forests are to be managed and used must take careful account of the economic and social environment in which they operate, so we review the basic structure of western economies, introduce ideas about society's fundamental economic goals, and sketch the role of governments and private producers. Finally, we outline a framework for policy development and decision-making, to indicate where economic analysis fits in.

APPROACHING FORESTRY FROM AN ECONOMIC VIEWPOINT

Forests are *economic* resources because we can use them to help produce goods and services that people want to consume. This is the definition of economic resources (or factors of production, as they are called in economics textbooks)—things in limited supply that can be combined with others to produce products and services that consumers want. Thus we can make use of a forest, combined with some labour and other inputs, to help produce consumer products like housing, newspapers, and outdoor recreation.

It is this usefulness of forests that makes them valuable economic resources. The more value in final goods and services that can be generated from a tract of forest the more valuable the forest itself.

Usually there is more than one way in which a forest can be used, and someone must choose among them. The timber might be harvested and used alternatively for making lumber, paper, or fuel. It might be kept standing, to support recreation or aesthetic values, or it might be saved for industrial use by future generations. Often, a forest can generate two or more kinds of benefit simultaneously, or sequentially—such as industrial timber, recreation, and livestock forage—in which case someone must choose the preferred combination and pattern of uses. In all cases choices must be made about how a forest will be managed, what goods and services will be produced, how much will be invested in enhancing growth, and so on. Economics is the study of such choices; specifically, the choices that determine how scarce factors of production are allocated among their alternative possible uses to produce useful goods and services. Forestry economics deals more narrowly with choices about how forests are managed and used, and how other factors of production like labour and capital are used in forest production.

Since forestry economics can be approached from several directions it is important to specify at the outset of this introductory book the viewpoint taken and some of the general assumptions underlying the discussion that follows. First, the focus of attention in this book is the forest land, the timber, and other goods and services produced directly from forests; it deals more with the primary resources of forest and land and less with the manufacturing and marketing of secondary forest products. It is thus concerned with the economics of natural resource management.

Second, our judgments about economic performance are made from the viewpoint of society rather than that of individual forest owners or producers. The criterion we adopt for assessing the economic advantage of one activity over another is a comparison of the net gain, the surplus of benefits over costs, that accrues to society as a whole, taking into account relevant concerns about the distribution of the benefits and costs. This is important, because the economic interests of individual entrepreneurs, landowners, or workers often diverge from that of society. In this respect this book differs from texts that take a business management approach and analyse problems from the narrower viewpoint of the producer or forest owner.

Third, we shall assume throughout that the purpose of forest management is to generate the maximum net value to society. This apparently obvious assumption is not insignificant, as much of the literature on forest management assumes, or at least implies, different objectives, such as production of the maximum possible quantity of wood, maximum profits to producers, or stability of harvest rates. Such concerns have an important place in forestry traditions and, as we shall see, they have had profound influence on forest policies in North America. It is important to recognize that narrower objectives of this kind inevitably conflict, to a greater or lesser extent, with the goal of maximizing the forest's economic contribution to society as a whole.

The value that a forest generates for a society can take a variety of forms. Some of these, such as industrial timber, are ordinarily marketed and their value is reflected in their market prices. Others, like aesthetic benefits and some forms of recreation, are usually provided free, so there is no market indicator of their value. Yet in assuming the viewpoint of society as a whole, we must take unbiased account of the full range of social benefits, whether they are priced or not. Much of our attention therefore will be directed to problems of evaluating environmental and other non-marketed benefits, tradeoffs among uses, and multiple use.

BASIC ECONOMIC QUESTIONS

An economy consists of production, consumption, investment, and other activities linked by a huge number and variety of transactions going on continuously. The bewildering detail and complexity of an economy can, however, be visualized in terms of a few straightforward processes.

On one hand, the society being served by the economy has certain wants. People *want* goods like food, houses, and television sets, and services like medical care and recreation. Their welfare or standard of living is measured by the extent to which these wants are met; the more people's wants are satisfied, the better off they are and, since no society has ever been known to be fully satiated, welfare is always a matter of degree. It is important to note that people's wants extend beyond strictly private desires to collective or public concerns about economic security, equity, and freedom.

On the other hand, any society has a limited capacity to produce the goods and services to satisfy these wants. The wherewithal consists of natural resources, man-made capital such as machines, roads, and other infrastructure, labour, and technical knowledge. All these change over time, but at any point in time they are finite.

The function of the economic process is to determine how these limited resources are used to satisfy some of the unlimited human wants. Thus economics is the study of how scarce resources are allocated among competing uses.

Every society must deal with three fundamental economic questions. Given its limited endowment of productive resources and the unlimited wants that they can serve, decisions must somehow be made about:

- *which goods and services,* of the almost infinite variety that it is technically possible to produce with these resources, will actually get produced, and in what quantities;
- *which* of the variety of technically possible *ways of producing* each good and service will be adopted in each case;
- *how* the goods and services produced will be *distributed* among members of the society.

These basic questions are answered in every economy, but in different ways. Primitive, subsistence societies make decisions about what to produce, and how, simply by tradition and custom. The socialist model relies on central planning and governmental direction. The capitalistic system depends on market forces generated by the un-coordinated actions of individual producers, consumers, and owners of productive resources.

Any study of the economics of forestry must take careful account of the character of the economic system within which forestry is being practised. A book on forestry economics in a socialist economy or in a subsistence economy must deal with quite different problems than one dealing with forest management under capitalism.

The typical form of economic organization in western industrial countries is "mixed capitalism," in which most production is organized and carried out by private entrepreneurs responding to market incentives. But governments play an important role in regulating economic activity, providing a variety of services, manipulating prices and incentives, redistributing income and wealth, and managing the general level of economic activity. This is the kind of economic organization we assume for the context of the discussion throughout this book.

The basic theory developed to explain how mixed capitalistic economies operate is thoroughly dealt with in numerous elementary textbooks. This book is intended to build on, rather than duplicate, this general economic theory. Accordingly, basic principles of economics are reviewed only briefly in the following chapters, emphasizing the particular role that forests play in the economic system and the economic choices faced by forest managers.

MIXED CAPITALISM AND THE ROLE OF GOVERNMENT

In a market economy entrepreneurs take responsibility for producing things and occupy the interface between suppliers of productive resources and purchasers of final goods and services. Entrepreneurs purchase the resources they need to produce the goods and services wanted by consumers, and the prices paid for these resources determine the income of those who provide resources. So the first of the three basic economic questions referred to earlier—what should be produced—is determined in the first instance by consumer demand, hence the concept of "consumer sovereignty." The second—about how the output will be produced—is determined by individual producers constantly competing to find the most cost effective means of production in order to enhance their profits. And the third—concerning the allocation of the fruits of production—is resolved by the distribution of income, which in turn is governed by the market values of the labour, capital, and other productive resources that private suppliers make available to producers.

But in the "mixed capitalistic" system typical of western countries, governments intervene in these processes. In important ways they influence the pattern of production. They not only provide the traditional "public goods" (such as roads, lighthouses, and national defence) that the market is not capable of providing, but also an increasing variety of goods and services that private markets can produce, but do so inadequately in the political judgment of the people. Such things as health care, education, and the arts fall into this category of "merit goods," which have a social value exceeding their value to the individual consumers of them. Moreover, western governments indirectly influence private production and consumption by means of taxes and subsidies. And governmental regulation of activities ranging from marketing to safety procedures for workers affect industrial structure, output, and prices. All these forms of intervention that alter the way productive resources are allocated and used comprise the *allocative* role of government.

The distribution of wealth and income is also substantially affected by modern western governments. Taxes, government spending programs, and transfers of various kinds all redistribute income among socioeconomic groups and regions. Sometimes these redistributional effects are deliberate and obvious, as when pensions are paid to the elderly, but often they are subtle and indirect, requiring complex analysis to trace their full impact. This is the *distributive* role of government.

Finally, modern governments accept responsibility for maintaining a stable level of economic activity. This calls for fiscal policies (spending and revenue-collecting programs) and monetary policies (manipulation of interest rates, exchange rates, and the supply of money) to offset trends toward inflation or unemployment. Related to this stabilization function are policies for promoting economic growth and regional development. These comprise the *stabilization* function of governments.

By intervening in various ways, governments attempt to correct some of the weaknesses and inadequacies of the market system. Expressed in other words, governmental intervention in the form of *allocative, distributive,* and *stabilization* measures reflects efforts to improve the performance of the economy in terms of achieving the economic objectives a society sets for itself through the political process.

In studying the economics of forest management we find ourselves continually confronted with governmental policies aimed at influencing the way forest resources are developed and used. The primary objective of some of these policies is to improve efficiency by affecting the rate and pattern of resource use. Other policies are motivated by distributional or equity considerations, or a desire to manipulate community and regional growth. But whatever their primary purpose, all forms of intervention inevitably have implications for all three of the fundamental forms of economic impact, namely the allocation of resources, the distribution of income and wealth, and economic stability and growth.

ECONOMIC OBJECTIVES: EFFICIENCY AND EQUITY

The allocative, distributive, and stabilization roles of government imply two fundamental economic objectives of society: *efficiency* and *equity*. These objectives provide us with criteria for assessing economic performance.

In any society that relies on market forces to guide economic activity there is a presumption, more or less qualified, that the primary objective of economic activity is to satisfy consumer demands to the greatest extent possible. The extent to which these demands are met with the available resources is a measure of the *efficiency* of the economic system.

At the macroeconomic level, if resources were employed in one sector of the economy when they could generate greater value in another, it would be possible through some reallocation to increase the value of total output and hence also the efficiency of the total system. In that case, the gross national product—the total value of all goods and services produced in an economy in a year—which is often used as a first approximation of an economy's performance, would be increased. Similarly, at the microeconomic level, if a producer fails to employ an available technology that would enable him to produce more with the same inputs, an inefficiency exists.

Efficiency thus refers to the relationship between inputs and outputs, and the greater the output relative to input the greater the efficiency. In economic analysis, efficiency is expressed as the ratio of benefits (outputs) to costs (inputs), both measured in the common denominator of dollar values. A thorough economic analysis from the viewpoint of society as a whole must, of course, account for non-priced benefits and costs as well as those that are more readily observed and measured in market prices.

Efficiency in economic activity is therefore a logical social objective. Unless there are offsetting considerations, the use of any resource in a way that generates less value than it is potentially capable of generating in some other use is simply a waste, lowering the value society derives from its resources.

How forests can best be used in light of the variety of demands on them is one of the central questions of economic efficiency in forestry. A second concerns the intensity of forestry; that is, how much labour and capital can be advantageously devoted to utilizing and managing forests to increase production. There is also an important temporal dimension to economic efficiency in forestry, referring to the pattern of investment and utilization of the resource over time. Because forests take so long to grow and can be harvested over such a wide span of time this temporal dimension of efficiency is especially important in forest economics.

Market economies give producers incentives to operate efficiently and thereby compete successfully. However, various distortions and market failures give scope for governments to improve efficiency through their allocative, stabilization, and growth-stimulating activities.

Equity implies some notion of a fair distribution of income and wealth, and therefore the fruits of production, among the population. As noted above, the distribution of income is determined, in the first instance, by payments for the factors of production. But it can be altered to a preferred pattern through taxes, transfers, and other distributive intervention by governments.

Like efficiency, distributive equity has more than one dimension. *Interpersonal equity* refers to the distribution of income among individuals at any time. Equity among people living in different geographical regions is referred to as *inter-regional equity*. And *intergenerational equity* refers to the distribution of income among people living at different times. All these dimensions of equity are relevant to forest policy.

Both efficiency and equity are difficult to measure. Efficiency is usually measured in dollar terms: the value of outputs relative to the cost of inputs, both of which are often reflected in market prices. But market prices are often misleading: some benefits are not traded in markets; some costs exceed the amount of compensation paid; and other distortions and market failures make it necessary to supplement market price information with estimates of social values in order to assess efficiency. Equity, which rests on subjective judgments about fairness in the distribution of income and wealth, defies empirical measurement except through political processes and ethical judgments.

It is important to note that the objectives of efficiency and equity often conflict and it becomes necessary to compromise one for another. For example, measures that could expand output (i.e., increase efficiency) might create unwanted changes in the distribution of income (i.e., decrease equity), and vice versa, illustrating the trade-off between improvement in equity and aggregate production and the choices that must be made. The relative priority of objectives and the appropriate compromises among them are not matters that can be solved by economic analysis. Political and electoral processes must be depended upon to prescribe the appropriate mixture of allocative, distributive, and stabilization efforts on the part of governments and to reconcile divergent opinion about equity and efficiency. Economic analysis can provide guidance in making these decisions, however.

It is important to note, also, that economic objectives cannot be pursued independently of other social objectives having to do with such matters as national security, protection of the natural environment, or cultural development. A society concerned with such issues is likely to find circumstances in which they conflict with purely economic objectives, calling for compromise at this level also.

In this book, heavy emphasis is put on the efficiency of resource allocation, especially the economic efficiency of forest resource development and use. This is not to suggest that concerns about equity and stability are unimportant in forestry; on the contrary, some of the most profound issues in forest management, which have motivated significant forms of governmental intervention, have to do with distribution among groups and regions and economic stability over time, as we shall see. But we emphasize the efficiency of resource allocation for two reasons. One is that it provides a necessary starting point for examining the effect on aggregate welfare of interventions aimed at affecting the distribution of income or economic stability and growth and vice versa. The second reason is that from the viewpoint of an economic analyst, much of the uniqueness of forestry, as distinct from other forms of economic activity, centres on problems of efficiency.

FORESTS AS ECONOMIC RESOURCES

In economics, the general term "resources" refers not only to land and natural resources but also to capital, labour, and human skills that are valuable in producing goods and services. The essential characteristic of an economic resource is that it is "scarce" in the sense that there is not enough of it available to satisfy all demands for it. It is this scarcity, or limitation of supply, that raises problems of choice about how resources are to be allocated. It also makes them valuable, even though their value in some uses is not reflected in market prices.

Not all forests are economic resources in this sense. Some are so inaccessible, remote, or poor in quality that they are not demanded for any economic purpose. Such forests, having no economic value or alternative uses, do not present the usual problems of choice and allocation among competing uses that are associated with economic resources. But most forests are capable of yielding one or more products or services, and so they constitute part of an economy's total endowment of productive resources. It is this economically valuable part of the total physical stock of forest that we are mainly concerned with in forestry economics.

An economy's total endowment of productive resources is commonly divided into four broad categories, namely *land*, *labour*, *capital*, and *entrepreneurship*. Each of these has distinctive economic characteristics, and each generates economic returns of a different kind, namely *rent*, *wages*, *interest*, and *profit* respectively.

Forest resources fall into two of these categories. The basic resource is the forest *land*, which has the same economic character as agricultural and other land. In any location it is fixed in supply; it varies in productive quality; and it generates a residual value, or rent, that varies accordingly. The *forest* itself, consisting of trees on the land, falls into the category of capital. It can be built up over time through investing in silviculture and pest control, or it can be depleted through harvesting; it derives its value mainly from final goods and services that can be produced from it; and it generates returns measured as interest. Standing timber is capital in this economic sense regardless of whether it is a gift of nature or a product of costly management.

Forest land and timber are economic resources because they are valuable in producing other, final goods and services. The demand for land and timber stems from the consumer demand for these final products, and in this sense is a "derived" demand.

Forest land, and the capital embodied in timber, are part of a society's total endowment of productive resources which can be used in a variety of ways to produce useful goods and services. Like other resources, the extent to which they contribute to social welfare is governed by the efficiency with which they are allocated and used.

Traditionally, forest economics has been concerned with the management of forests for production of wood for industrial manufacture into building materials, pulp and paper and so on. But forests also yield other goods and services and are often managed to produce livestock, fish and wildlife, recreation, and water supplies. Such benefits are often produced in combinations with industrial timber. Some of these values, especially recreational and environmental benefits, have become increasingly important in recent years.

These increasing and overlapping demands on forest resources complicate the problem of allocating them among alternative uses and combinations of uses. Moreover, while there is usually a market price to indicate the economic value of industrial timber, some other benefits such as outdoor recreation and aesthetic and wildlife values are often not priced. This leaves them difficult to evaluate in terms comparable with timber values. But their value is real whether they are priced or not; the absence of price indicators only complicates the problem of economic analysis. These are issues addressed in detail in later chapters.

FORESTRY AS APPLIED SCIENCE AND APPLIED ECONOMICS

Economics deals with all kinds of productive resources, while forest economics focuses specifically on those used in forestry. The latter includes, obviously, the land and forest growth that constitute the forest itself. But it must also consider the labour, capital, and other inputs to forest operations. Much of forest economics is concerned with how much of these other resources can be efficiently combined with forest land and timber in producing forest products and services.

This is the subject matter of microeconomics—that half of economic science that deals with how prices and incomes are determined, how producers find the most efficient scale and form of production, how consumers behave, and so on. Forest economics builds on this basic theory as it applies to forests. Forest economics is thus, in large part, a study in applied microeconomics.

Like other special fields of applied economics, forest economics draws on the particular threads of economic theory that are relevant to the unique or especially important problems of the field. For forest economics, the theory of production, and especially the theory of capital and rent, are fundamental. And, as a relatively narrow area of applied economics, it draws on broader applied fields such as the long-established specializations in land and agricultural economics and the newer branch of natural resource economics.

The special characteristics of forest resources, which justify forest economics as a special field of study, can be summarized as follows:

- Forests can produce a wide variety of goods and services and combinations of them, some of which are not priced in markets. This gives rise to special problems relating to the allocation of resources among uses.
- Timber is an unusually slow-growing crop, often involving in-

vestment periods of many decades. This gives rise to special problems in analysing investments, harvesting schedules, risks in carrying forest crops over long periods, and market uncertainty. It also means that forests can be altered only slowly in response to changed economic conditions.

- Forestry usually involves very high capital and carrying costs relative to production because the slow rate of forest growth means that large forest inventories must be carried to sustain a modest harvest. As a result, the costs of forest production are often dominated by the burden of carrying land and capital over time.
- Forests valued for industrial timber are both productive capital and product. This fact distinguishes forests from other forms of capital and gives rise to special analytical problems in selecting the best age to harvest and in designing taxes and regulations.

These features are not unique to forestry, but forestry illustrates them in a unique degree. And they are issues which underlie most of the analytical problems addressed in this book.

The economic choices that can be made in forest management are constrained by the biological capacity of the resource. Those limits, and the scope for manipulating them, are the subject of the natural science of forestry, or silviculture. Silviculture is a specialized field of biology, just as forestry economics is a specialized applied field of economics. And while silviculture is concerned with all the things that can be done to manipulate the structure and growth of forests, forest economics deals with decisions and choices within that range of possibilities, focusing attention on their social, rather than their biological, implications.

However, forestry economics is concerned not only with silviculture but with all aspects of forest management—protection, development, harvesting, and utilization of the full range of goods and services associated with forests. The natural science of forestry identifies the limits of natural systems and the range of choices available to forest managers; this range provides the framework of natural constraints within which economic analysis can help in identifying the social implications of alternative courses of action.

ECONOMIC DECISION-MAKING

Economic activity can be viewed as a process of decision-making, and decisions about how forests are managed and used can be viewed as economic decisions about the allocation of resources in the broad sense. Decision-making about forests is thus part of the much larger mosaic of resource allocation activity that goes on continuously in an economic system.

In mixed capitalistic economies most of these decisions are made by private firms and individuals pursuing their economic self-interest but, as noted above, they are influenced and constrained by decisions of governments. Forest economics is concerned, in large part, with how decisions about forests are made, how effective the decisions are in enhancing economic performance and social well-being, and how the decision-making can be improved.

Forest resource management ranges from the design and implementation of high policy to the execution of everyday field tasks. The person or body that makes the decisions varies according to the allocation of responsibilities. Broad policy objectives are determined by governments and corporate boards of directors; how particular forests are to be used is usually the decision of their private or public owners; for detailed matters it is often foresters, superintendents, or foremen employed by the owner who make the decisions.

Whatever the level, the process of decision-making can be viewed as involving at least the following five steps: identification of goals or objectives, identification of the alternative possible means of pursuing those objectives, evaluation of the alternatives, choice of the preferred alternative, and implementation of the decision. In practice, decision-making seldom follows the orderly sequence implied by this list of steps. The objectives of those involved are often unclear or conflicting, their motives may range from self-interest to altruism, and their time perspective may range from the immediate to the distant future. The processes of investigation and evaluation often bring out new information that causes those involved to change positions and shift alliances. As a result of this ongoing process, decision-making often appears confused and disorderly, especially in matters of public policy. Nevertheless, it is helpful to the understanding of decision-making to identify these separate components of the process.

Objectives

To make appropriate decisions, the decision-maker needs a clear purpose to serve as a frame of reference from which he can judge the desirability of one course of action over another. Thus a forest manager facing a decision about how to plan a harvesting program, or how much provision should be made for wildlife, or where to direct silvicultural effort, must assess his alternatives in light of the objectives he is striving to achieve. He must know the object of the exercise.

Several points about objectives deserve mention. First, objectives, even at the same level of decision-making, vary depending upon who is responsible for defining them. A government, for example, is likely to have different and broader objectives for the management of public forests than a corporation for the management of its private forest land. A major objective guiding the forest operations of industrial corporations is their responsibility to shareholders to generate profits, and corporate decisions are typically directed toward increasing profits. But they may be influenced by other goals as well, such as corporate growth, security of markets or resource supplies, or protection of dependent manufacturing activities, and they may be prepared to compromise their profit earnings to advance these other goals. Small private landowners may similarly be guided by desires not only for profit but also for financial security, amenity, or the prestige they can derive from their forests. The management of public forests in democratic countries reflects the perceived wishes of the populace, which nowadays typically puts considerable emphasis on the non-marketed and environmental benefits of forests, on distributional considerations, and on regional development. In short, those who make the decisions about how forests are to be managed have varying frames of reference and hence differing objectives that lead to differing decisions.

Second, the objectives of decision-makers depend on the hierarchical structure of the organizations within which they work. As one moves down through the organizational structure of a government or corporation the relevant objectives of decision-makers become more narrowly defined. For example, at the highest policy-making level in a government, the goal might be to promote regional economic stability. Toward this end, the government's forest management agency might adopt a sustainable yield objective in regulating timber harvests in each region. That objective would provide regional administrators with production objectives, the official in charge of silviculture with reforestation objectives, and the foreman of the planting crew with daily planting targets. This example illustrates that at each subsidiary level of decision-making the objective is different and narrower, but derives from and is consistent with the next higher objective and ultimately with the general goal of advancing regional economic stability.

It is important to distinguish ends from means in this context, because they are often confused. An example is sustained yield (examined in Chapter 8); this is a principle that has become so enshrined in the forestry administration of some jurisdictions that it has become institutionalized as an end in itself. But it is merely a formula for meeting a higher purpose, such as regional industrial stability, and unless it is clearly seen as a means to such an end its limitations for that purpose, and the implications of alternatives to it, cannot be properly assessed.

Third, forest managers are often expected to pursue several objectives simultaneously. As suggested already, a corporation might be concerned with such things as security of raw material supply or avoidance of risk as well as profit maximization, and a government may seek to provide stable regional employment or environmental benefits as well as revenues from a public forest. These various objectives are rarely perfectly complementary, and so to pursue them together requires compromises among them. Economic analysis can assist in identifying and evaluating possible trade-offs, but the ultimate choice usually requires some weighting of the competing values which are often not easily quantifiable, as discussed later.

Fourth, while orderly decision-making calls for explicit objectives, the objectives that forest managers are intended to pursue are sometimes vague. This is a difficulty faced most frequently in governmental forest agencies, where guidance about the broad purposes to be served in managing public forests is often ambiguous or even conflicting in the legislation, regulations, and administrative arrangements that articulate public policy. In these circumstances managers are forced to infer, or guess, about their intended objectives, which can lead to inconsistencies and inefficiencies.

Finally, it is worth emphasizing that specification of objectives is not the responsibility of an economist. The expertise of the economist is not in prescribing corporate or governmental goals but, as discussed below, in analysing and evaluating the means of achieving them.

Identifying Alternative Means

Most corporate or public goals can be pursued in a variety of ways. At the level of high economic policy, a goal such as increased regional employment might be served by promoting industrial development, for example, and there are many means of doing this through taxes, subsidies, infrastructural improvements, or direct governmental enterprise. Forestry may be only one of several opportunities. Or, at the level of forest management planning, a goal of increasing yields might be accomplished by such varied means as improved protection, reforestation, spacing and fertilization of stands, or closer utilization of harvested trees. Thus once the decision-maker's objectives are identified, the next step is to identify the range of alternative strategies that can feasibly be adopted to serve those objectives.

This step involves assessing the technical alternatives and the inputs required to achieve a particular level of output, which in economic jargon is referred to as determining the production function. Sometimes it is important also to assess the risk or uncertainty associated with the alternative strategies.

Identification of the feasible means of pursuing an objective, and their technical production functions, is typically the responsibility of engineers, foresters, and other technical experts. For large ventures, this task sometimes becomes highly formalized in feasibility studies, while at the operational field level it typically depends on continuing subjective assessments by working supervisors.

Evaluation

The next step is to evaluate the technical alternatives to provide guidance in choosing among them. It is at this stage that economic analysis is brought to bear on the decision process. It involves assessing the extent to which the goals of the decision-maker would be advanced by a particular action and the costs of doing it.

The relationship between the value of the output and the cost of the inputs associated with a particular activity provides a measure of the potential net gain it can generate. Economic efficiency calls for maximizing the surplus of benefits over the cost of resources utilized, so the greater the value of output relative to the cost of inputs, the more efficient is the activity. Chapter 6 describes how alternative courses of action can be assessed according to efficiency criteria.

The task of identifying the relative advantage of alternative courses of action is often complicated by incomplete information, distorted or non-priced costs and benefits, and uncertainty about future circumstances and outcomes. Notwithstanding these difficulties (which are examined in subsequent chapters) economic evaluations offer the means of ranking alternative courses of action according to consistent criteria for the guidance of decision-makers in selecting among the alternative strategies available to them.

Choice and Implementation

The economist's role in evaluating alternative courses of action is to provide guidance in decision-making; it is not to make the final choice. Decision-makers may, for a variety of reasons, reject the activity that appears most advantageous on purely economic grounds because of considerations of corporate strategy, political sensitivities, or other concerns not accounted for in the analysis. Nevertheless, economic analysis will assist decision-makers to better understand the implications of their choices.

The final step in decision-making is to initiate the course of action decided upon, a procedure which depends on the nature of the issue at hand. Sometimes an additional step is added, that of review and assessment of the action taken, drawing attention to the dynamic and continuous character of decision-making and the need for evaluation.

DECISION-MAKING AND ECONOMIC ANALYSIS

Decision-making can be regarded as the interplay of objectives, courses of action, and outcomes. The *objectives* of the decision-maker provide the motive for taking some action and the basis for assessing results. The *courses of action* are the alternative measures that he can take. And the *outcomes* are the results of each alternative action. Economic analysis involves evaluating the outcomes of alternative actions with reference to the objectives.

Economic decisions are never made with complete certainty, of course. Information about the resources and markets, the range of possible actions and their outcomes is always more or less uncertain. Most decision-makers are averse to risk, and so the degree of uncertainty surrounding alternative courses of action is a significant influence on their choice. But attitudes toward risk-taking vary considerably.

The degree of uncertainty is therefore an important dimension of decision-making, and a later chapter considers how it can be taken account of in economic analysis. It is particularly important in forestry because of the usually limited knowledge about the biological character of forests and their potential responses to treatments. As well, the long planning periods involved in forest production aggravate the difficulty of predicting the future values of products and services which will determine the economic outcome of current actions. Risks of losses from fire and other causes also contribute to the uncertainty in forestry decision-making.

Economists traditionally have approached their subject in two ways. "Positive" economics is concerned with describing and explaining economic behaviour, without judgments about its desirability; in contrast, "normative" economics assesses behaviour in terms of given criteria or objectives and is therefore more concerned with how economies *should* be organized and regulated. This introduction to forest economics does not follow either of these schools exclusively. Students of forest economics are concerned with understanding the economic process as it bears on the use of forests, but they are also interested in using economic analysis to assist in making decisions. Their main purpose is to apply economics to forestry problems to help identify desirable courses of action in light of private or public objectives.

The process of decision-making sketched in this chapter is a central concern of management science. The growing literature on this field of study presents a variety of decision models and models of strategic behaviour to help understand the relationships among decision-makers, the problems of multiple and conflicting objectives, means of minimizing and coping with uncertainty, and so on. Economic analysis contributes to decision-making processes by providing guidance to decision-makers.

The following chapter reviews the forces that guide decision-making in the context of a market economy. Subsequent chapters examine the special problems encountered in forest management and how economic analysis can contribute to their solution.

REVIEW QUESTIONS

1 Explain why economics is concerned only with the allocation of "scarce" resources. In what sense are forest resources "scarce"?
2 Compare how management decisions are made for (a) a privately owned forest in a capitalistic economy, (b) a government-owned forest in a planned socialistic economy, and (c) a tribal forest in a primitive subsistence economy.
3 Explain how innovations in mechanized forestry can affect (a) economic efficiency in timber production, and (b) the distribution of income.
4 Describe the importance of objectives in evaluating forest management decisions.
5 Compare the approaches of a silviculturalist and an economist in considering how best to manage a forest. What are the main concerns of each likely to be? Can their approaches be reconciled?

FURTHER READING

Clawson, Marion. 1975. *Forests for Whom and for What?* Baltimore: Johns Hopkins University Press for Resources for the Future. Chapters 3 and 7

Duerr, William A. 1988. *Forestry Economics as Problem Solving*. [Blacksburg, VA]: By the author. Part 1

—, Dennis E. Teeguarden, Neils B. Christiansen, and Sam Guttenberg (eds.). 1982. *Forest Resource Management: Decision-making Principles and Cases.* Corvallis, OR: OSU Book Stores. Part 2

Gregory, G. Robinson. 1987. *Resource Economics for Foresters*. New York: John Wiley & Sons. Chapter 1

Howe, Charles W. 1979. *Natural Resource Economics: Issues, Analysis, and Policy*. New York: John Wiley & Sons. Chapter 1

Quade, E.S. 1982. *Analysis for Public Decisions.* 2nd ed. New York: North-Holland. Chapters 4–7

Rumsey, Fay, and William A. Duerr (eds.). 1975. *Social Sciences in Forestry: A Book of Readings*. Philadelphia: W.B. Saunders. Items 1, 2, and 12

Samuelson, Paul A., and William D. Nordhaus. 1989. *Economics*. 13th ed. New York: McGraw-Hill. Chapters 1–3 and 32; or, for Canadian students: Samuelson, Paul A., William D. Nordhaus, and John McCallum. 1988. *Economics*. 6th Canadian ed. Toronto: McGraw-Hill Ryerson. Chapters 1–3 and 32

CHAPTER TWO

Economic Efficiency, Market Processes, and Market Limitations

Out in the Forest . . .

The continuing efforts of Peavey Forest Products Limited to generate profits require the company's managers to keep searching for more efficient ways of doing things. This means getting more production for the same inputs, or the same production with fewer inputs. The company's "bottom line" of profitability shows up in its financial statements as the difference between the revenues it receives from the products it produces and the costs it incurs in producing them.

David Cameron, the company president, explores all possibilities to improve efficiency. He insists that his forester, Ian Olson, compare the returns to different kinds of silvicultural treatments on different parts of the forest so that funds can be allocated to their most beneficial uses. Olson also evaluates all opportunities to buy more forest land in the area, and compares the cost of increasing the company's timber supply this way with the alternative of investing in silviculture. The company's controller evaluates the benefits of buying new equipment, weighing the capital cost against expected gains in labour productivity and lower maintenance costs. In consultation with his log sales manager Cameron assesses price trends and weighs the advantages of increasing present output at the expense of future production and vice versa. The search for improvements in efficiency goes on continuously and in all phases of the company's operations.

If Peavey Forest Products Limited and all other producers in the economy succeeded in achieving the maximum possible efficiency in production it might be expected that the whole economy would correspondingly achieve maximum efficiency, so that the greatest possible value of goods and services would be produced given the resources available. But there are many obstacles to such perfection. A company like Peavey Forest Products Limited often finds the markets in which it sells its products and buys its inputs are not very competitive; sometimes it inflicts costs on others, such as by impairing the amenity of the landscape by logging, which do not enter its own cost accounting; occasionally financial pressures force the company to harvest faster than the long-term interest of society would warrant; and sometimes it just makes mistakes.

So although David Cameron and his associates manage Peavey Forest Products Limited as a relatively efficient and successful company, it is not perfect. Nor is the economic environment in which it operates. Governmental efforts to correct the resulting distortions explain many of the controls and regulations on the company's operations.

Decisions made by forest managers can be viewed as economic decisions involving the allocation of resources in the broad sense. The focus of immediate interest is typically land and timber, but decisions about these also involve decisions about the use of other resources such as labour, capital, and entrepreneurial skills. In effect, forest management decisions determine how, and how much, of a wide variety of economic resources will be directed to forest production. Forestry decision-making is thus part of the much larger mosaic of resource allocation activity that goes on continuously in an economic system.

This chapter deals with the efficiency of economic decision-making and the deviations from efficient results we frequently observe in forestry. It begins with a brief outline of the economic theory of production. We then review the conditions that must be met in a market economy to ensure that all resources will be used with maximum efficiency. This review provides a framework for examining how the markets affecting forestry decision-making often fall short of the perfectly competitive market conditions that promote economic efficiency. These so-called "market imperfections" or "market failures" give rise to the special economic problems in forestry that provide much of the substance of forest economics as a field of study.

Pulling together the relevant elements of economic theory is a necessary precursor to later chapters, but it must be said at the outset that it is a rather abstract and mechanical task. Readers already familiar with microeconomics may want to pass over this chapter quickly.

ECONOMIC EFFICIENCY AND OPPORTUNITY COSTS

Economic analysis is based on the plausible presumption that the objective of all economic activity is to satisfy human wants, and the greater the total satisfaction achieved with the limited resources available, the greater the efficiency of the economic system.

To determine whether productive resources are being used efficiently, we must compare their productivity in their actual use with the alternatives. For example, if a parcel of land—or a unit of labour or capital employed in timber production—could otherwise produce agricultural products, the cost to society of employing them in forestry is the value of agricultural output foregone by doing so. This is the *opportunity cost* of factors of production: the value of output sacrificed by not directing them to their best alternative use.

Where markets work perfectly, the opportunity costs of inputs are accurately reflected in their prices. The wage paid to truck drivers, for example, will tend to reflect their marginal productivity in the various industries that employ them. But sometimes prices are unreliable guides to opportunity costs. For example, if some of the labour employed in a forestry operation would otherwise be unemployed, then there is no sacrifice in other production by employing it; its opportunity cost is zero even though its price (or wage) is positive.

Opportunity cost measures the *real* cost to society of using a resource in a particular way. However, private producers in their pursuit of profit respond mainly to the *monetary* cost or price they must pay for their inputs. As a result, whenever the market prices of inputs differ from their opportunity costs, an assessment of the efficiency of an economic process by a private corporation will diverge from that of an analyst taking the viewpoint of society as a whole (as is done here). Both compare benefits with costs, but the strictly monetary benefits and costs that private producers use to assess profitability must often be supplemented or adjusted to estimate more accurately the net gains to society as a whole.

THE THEORY OF PRODUCTION

There are always alternative ways of producing things; the task is to find the most efficient, which means that every resource is employed in its most productive use. For example, timber can be harvested using different combinations of machinery and manpower, each having a different opportunity cost. The combination that can harvest the timber at the lowest cost is the most efficient. And if the market prices of these inputs accurately reflect their opportunity costs, then the combination chosen by private producers will converge with the most efficient combination from the viewpoint of society as a whole. The concept of economic efficiency thus focuses attention on the relationship between production and the cost of the resources used in the production process—the relationship between inputs and outputs.

The Appendix to this chapter reviews the basic concepts underlying the economic theory of efficient production, and they are examined in much greater detail in textbooks on microeconomics. The following paragraphs only summarize the essential relationships referred to later.

Inputs, outputs, and substitutability. The theory of production begins with the technical relationships between inputs and outputs; that is, how technology enables producers to combine inputs in various ways to produce particular products. The relationship between the inputs in a production process and the output is referred to as a *production function*. A production function for timber, for example, might indicate the relationship between the output of timber and varying inputs of land and labour.

It is almost always possible to substitute one input in a production process for another. For example, production of timber requires inputs of land and labour, but it is possible to grow the same amount of timber using more land and less labour and vice versa. The degree of substitutability among inputs depends on the technology of production. Because technology advances over time, so does the scope for substitution among inputs in production.

As more of one input is substituted for another, it takes larger and larger increments of it to maintain the same level of output. This illustrates the *law of diminishing marginal substitutability*, which explains how it becomes increasingly difficult to substitute one input for another while maintaining the level of output.

Least-cost combination of inputs. Given the substitutability of inputs in a production process, and hence the variety of possible combinations of them that can serve to produce a given quantity of product, the most efficient combination is that which entails least cost. This calls for information about the cost, or price, of inputs as well as their substitutability in production. The least-cost combination of inputs to produce a given output is precisely defined as that at which the marginal rate of substitution of one input for another is equal to the ratio of their costs.

Efficient level of output. Such a least-cost combination of inputs exists for any level of production, so it remains to identify the most efficient level of output. In most production processes, efficiency depends partly on the scale of operations. Beyond some point, increased inputs will not result in proportional increases in output, manifesting *decreasing returns to scale*. This means that additional increments in output will require larger and larger additions of inputs, raising the producer's marginal cost; that is, the cost of producing another unit of product.

Producers will increase their profits by expanding production as long as the additional revenue they receive by producing and selling another product exceeds their marginal cost of producing it, but not beyond that point. Thus producers seeking maximum profits select the level of production at which their *marginal revenue* equals their *marginal cost*.

This profit-maximizing behaviour explains why producers are almost always willing to produce and sell more at higher prices, as reflected in upward-sloping market supply curves for products. At a higher price each producer must expand production until his marginal cost rises to the level of his new marginal revenue.

It also explains why producers will demand more inputs the lower their prices, as reflected in downward-sloping demand curves. Lower input prices mean lower marginal cost, so producers must expand production, using more inputs, until their marginal costs rise to equal their marginal revenue.

Time as an input in forest production. The relationships outlined above help to explain how a profit-maximizing producer chooses his level of output and his inputs at a point in time, and the behaviour of forest-producing enterprises can be analysed satisfactorily within this theoretical framework. But in producing forest crops, producers must pay special attention to time, in addition to the usual kinds of inputs that must be considered in manufacturing and other forms of production. Once a forest is established, the time it is left to grow is often the most important determinant of the output produced.

Time is therefore one of the variable inputs in forest production, and it has the essential characteristics of other inputs. Generally, the more time a managed forest crop is left to grow, the greater will be the output; the volume of output shows diminishing returns to time; and time involves a cost, measured by the opportunity cost of tying up the capital and land as long as harvesting is postponed. Moreover, because the same output can be produced in less time with the use of more labour and the other inputs of intensive silviculture, the optimum combination of inputs must take account of this substitutability between time and the other factors of production. Chapter 6 deals specifically with time as a special dimension in the economics of forest production.

FACTOR ALLOCATION AMONG USES

Most factors of production can be used for a variety of purposes. Land used in producing timber can often be used for agriculture, recreation, or urban development. Similarly, the labour and capital employed can alternatively be used in producing other things. We must now examine the conditions for efficiently allocating factors of production among the various productive purposes for which they can be used, as well as among all the enterprises producing similar products.

An additional unit of one of the inputs employed in a production process will increase output by an amount referred to as its marginal product or, more precisely, its *marginal physical product* (MPP). If all the other inputs are held constant while one input is increased by successive equal increments, output will increase, but the resulting increments in output will become smaller and smaller. For example, the more labour devoted to cultivating the timber on a tract of forest land the more timber can be produced, but the increase will not be proportional; the more labour applied the less an additional unit of labour will contribute to timber production. This is the important law of diminishing marginal product, more commonly referred to as the *law of diminishing returns*.

The *value* of the additional output generated by adding another unit of an input is the *marginal revenue product* (MRP) of the input. It is the marginal physical product of the input multiplied by the marginal revenue the extra product yields. The marginal revenue product of an input declines as more of it is employed because diminishing returns causes the marginal product to decline and, in imperfectly competitive product markets, because the marginal revenue declines as well.

A profit-maximizing producer will employ more of any factor as long as its marginal revenue product exceeds its cost, because this will contribute to profit. The same is true for producers of other products that use the same input. As a result, all the producers purchasing the input in a competitive market will bid its price up to a single level which will reflect its marginal revenue product in all uses. When the marginal revenue product of a factor of production is thus equal in all its uses in an economy it can be said to be efficiently allocated. This is because it will then be generating a value at least equal to its opportunity cost in all uses, and there can be no possible reallocation that would increase the aggregate value of production.

The same principle applies to a firm producing several products, in deciding how to allocate an input efficiently among different forms of production. For example, manufacturers of timber often face the task of allocating logs among several uses such as lumber, pulp, and paper production. Again, maximum efficiency can be gained by allocating the logs in such a way that their marginal revenue product is equal in each use, and equal to the price of logs.

These conclusions can be generalized to describe the conditions that must hold throughout the economy in order to achieve maximum economic efficiency, or the greatest possible benefit to society from all the resources available for production. If the revenue that producers earn from employing each factor of production, at the margin, accurately reflects the value of society of the output it contributes, then the marginal revenue product measures its *marginal social benefit* (MSB). And if all producers employ each factor to the point at which its marginal revenue product equals its price, that price will reflect its true opportunity cost, or its *marginal social cost* (MSC). Resources will be allocated to achieve maximum social efficiency if

$$MSB = MSC$$

in all forms of production. Then no reallocation could yield a higher social value in the aggregate.

MARKET EFFICIENCY AND MARKET FAILURES IN FORESTY

Private producers and market processes can achieve this socially optimum result only under the stringent conditions of a "perfect" market economy. These conditions are never fully met, of course; in reality market processes are always more or less imperfect. Nevertheless, the theoretical model of a perfect market system helps to identify the sources and effects of imperfections in the real markets we have to deal with and the opportunities for improving their performance.

Why can we not simply leave the business of forestry to the free workings of the market economy and expect satisfactory results? This is another way of asking what are the weaknesses and failures in market processes that impair the efficient production and use of forests in market economies. The following paragraphs summarize the conditions that must prevail in a market economy in order for it to achieve the socially optimum allocation of resources identified above, and some of the ways in which these conditions are not met in forestry.

Profit Maximization

A fundamental precept of the market system is that producers seek to maximize their profits, and the efficiency of the system depends on them doing so. As we have seen, this incentive will lead them to choose the most efficient or least-cost method of production and to select the level of output at which their marginal revenue equals their marginal cost (MR = MC).

For purposes later in this chapter, it is convenient to cast this profit-maximizing rule in terms of an equality between the marginal cost and revenue associated with employing another unit of a factor of production (rather than those associated with one more unit of output). Then the rule is that the marginal revenue product of the factor (MRP, which we defined earlier as the factor's marginal physical product multiplied by the producer's marginal revenue) is equal to the cost to the producer of employing the additional factor, the marginal factor cost (MFC); that is

$$MRP = MFC.$$

If the firm fails to maximize its profit in this way, it will misallocate resources in the sense that they could be used more efficiently in some other way. For example, if a firm producing timber produces beyond this point of equality of marginal revenue and marginal cost, the cost of producing the last cubic metre of timber will be greater than the revenue it generates. This not only reduces the firm's profits but also means that the resources used in producing the timber, at the margin, are capable of generating more value in other things.

There are many ways in which forest production and use fails to be guided by profit maximization. Some forests are owned not by private firms but by governments that often pursue other objectives. Even private owners are constrained in their profit-maximizing behaviour by governmental regulations. Some forest products and services, like recreation, amenity, and other environmental values are not marketed and their value cannot be realized by private forest enterprises. Such circumstances prevent forest production from being consistently guided by the marginal costs and revenues incurred by private firms, and thereby impede efficient allocation of resources through market processes.

Perfectly Competitive Product Markets

To provide producers with incentives to produce at the socially most advantageous level, the markets for the goods and services they produce must be perfectly competitive. *Perfect competition* is characterized by a large number of sellers of a homogeneous product, each seller being too small relative to the total market supply to influence the product's market price.

Such market conditions imply that the firm's marginal revenue, the extra revenue it gains from selling another unit of product, must be equal to the price of the product. It follows that if the firm alters the amount it sells, the change in its revenue will be equal to the market price multiplied by the change in quantity sold, and that this will be identical with the market's evaluation of the change in output. In other words, the market value of the marginal product (VMP) must equal the firm's marginal revenue product (MRP), i.e.:

$$VMP = MRP.$$

This condition will not hold if, as is often the case in forest product markets, competition is imperfect. For example, when a firm has a monopoly in a market, or is such a dominant seller that it can influence the market price of its product by adjusting how much it sells, its marginal revenue will be less than the price of its product. Moreover, when such a firm equates its marginal cost and marginal revenue to maximize profits, its marginal cost will be less than the price of the product, giving rise to another inefficiency.

Relatively unrestricted international trade in softwood lumber, with large numbers of buyers and sellers of fairly homogeneous products, creates market conditions that approach the perfectly competitive model. But regional markets for raw timber, pulp, newsprint, plywood, and other intermediate products are rarely perfectly competitive. Some have the character of *oligopolies*, where the market is dominated by only a few sellers, each with enough market power to set its own prices, within limits, and to influence the market opportunities of the others. Extreme cases of the absence of competition are *monopolies*, where there is only one seller. *Oligopsonies* and *monopsonies* describe corresponding limitations to competition on the side of the demanders for a product.

A condition for perfect competition is the absence of barriers to entry. Significant barriers to the entry of new producers in markets for some forest products such as pulp and paper arise from the large economies of scale and heavy capital requirements of these industries. Moreover, because forests grow only slowly, large timber-producing enterprises require extensive areas of operation which, once established, can restrict opportunities for new ventures to gain access to raw material supplies.

Finally, competition may be impeded by integration of forest products enterprises. The diversity of type, size, and quality of timber recovered from most forests, and indeed from individual trees, means that the raw material harvested can best be utilized in a variety of products. This encourages horizontal integration among enterprises producing such products as lumber, plywood, and pulp and paper. Moreover, the substantial capital investments required for forest products manufacturing encourages producers to vertically integrate in efforts to secure reliable sources of raw material on the one hand and markets for their products on the other. Highly integrated forest industries narrow the scope for competition in intermediate product markets.

Perfectly Competitive Factor Markets

Correspondingly, the efficient operation of the market system requires that the markets for all factors of production are perfectly competitive, so that no single producer or factor owner can influence factor prices. Thus each producer must pay the going market price for all of the factors he purchases. He is a price-taker, facing horizontal factor supply curves; and correspondingly suppliers face horizontal demand curves for their factors at the going market prices.

Under these circumstances, the producer's cost of producing an additional product, his marginal cost, is equal to the value of the factors needed to produce it. Put another way, the cost to him of employing another factor, the marginal factor cost (MFC) must be equal to the price he pays for it, the value of the marginal factor (VMF), i.e.:

$$MFC = VMF.$$

This condition will not hold if those who supply factors have monopolistic power or if those who purchase them can exercise monopsonistic powers to influence factor prices. Monopsonies in local land and timber markets, bilateral monopoly bargaining in organized labour markets, and restricted access to capital markets for small landowners are examples of impediments to competition in the markets for factors of production.

Divisible Inputs and Outputs

The optimum allocation of resources requires fine marginal adjustments in production decisions, which implies that all inputs and outputs are highly divisible. In natural resource activities, however, the scale with which inputs are used is often constrained by natural circumstances. The "natural" management unit for an industrial forest may be a drainage basin or forest type that does not readily lend itself to division. A wilderness usually requires a significant area to maintain its natural character. So while the theory of market economies rests on discrete, divisible, homogeneous, and mobile inputs, the natural resources used in forest production are often heterogeneous, immobile, multipurpose, and sometimes, indivisible, restricting producers' flexibility in making adjustments at the margin.

Control of Inputs

An efficient market system must provide each producer with complete control over his inputs and outputs. He must be able to acquire the inputs he needs, to use them in the most efficient way, and to dispose of the goods and services he produces in the most beneficial manner, incurring the full benefits and costs of his actions.

Enterprises that use land and natural resources rarely have full control of the beneficial use of them. An obvious impediment, more common in fisheries and water management than forestry, are common property regimes where producers are never free of interference from other users of the same natural resources. In these circumstances, producers lack appropriate incentives to manage and conserve because the benefits will accrue, in large part, to others. But even where producers acquire exclusive rights to resources, governmental taxes, royalties, and other charges drive a wedge between the private and social benefits of using them. As Chapter 10 will show, these levies may create incentives to use resources in ways that will reduce the social value they generate.

A wide variety of leases and licences provide forest enterprises with only qualified property rights in land and timber, described in Chapter 9. They typically limit the holder's rights to a particular resource, such as timber, while excluding rights to others, such as the water or wildlife on the same land; they usually involve charges that distort incentives governing the way he manages and uses the resources; and they are of limited duration, which may lead the holder to ignore the long-term impact of his actions.

In short, while the theory of the market system rests on well-defined private ownership of all factors of production, private property in forest resources is often absent or constrained. Limitations on private ownership and control may serve important social objectives, which we consider in later chapters; here it is important to note that they distort the incentives of operators and owners and thereby interfere with the efficient performance of market economies.

Optimal Time Preference

Forest management, like many other economic activities, involves decisions about timing: when to harvest a stand, when to apply silvicultural treatments, when to invest in roads, and so on. Forest planning also calls for ways of evaluating investments that yield returns only over long periods. The key to comparing values that occur at different times, discussed in Chapter 6, is the rate of interest, which measures the degree by which present values are preferred over future values.

In a market economy, decisions about the relative value of a dollar today and a dollar at some future date are made by private demanders and suppliers of capital with reference to the market rate of interest. Thus, in order for the market to allocate resources over time in the best interest of society as a whole, the rate of interest used by private producers in making their decisions must accurately reflect the preference of present over future values for the society as a whole. In other words, market rates of interest must reflect *the social rate of time preference*.

For many reasons, this may not be the case. All producers rarely have equal access to capital markets, so that even if some market rates correspond to the social rate of time preference others do not. For example, if small forest owners have to pay higher interest rates on loans than large corporate owners of similar forests, they will be less willing to invest and more likely to discount future values at a higher rate than society as a whole. Moreover, as we shall see later, the broader interests of society may mean that market rates of interest are an unreliable guide to social time preference.

No Externalities

For the market system to work perfectly, those who make production decisions must take account of all the costs and benefits of their actions. However, many forest activities involve *external economies* and *diseconomies* which are beneficial or adverse impacts that do not enter into the economic decisions of private producers because they do not involve market transactions.

Forests are capable of producing a variety of goods and services simultaneously, and the exploitation of one often affects the availability of others. Timber harvesting may enhance the habitat for some wildlife species, for example, or it might diminish recreational opportunities, cause pollution, or impair the aesthetic quality of the landscape. Externalities are the market imperfections that occur whenever such impacts are not compensated through markets: in this case, where the logging enterprise that generates some beneficial side-effect is not paid for doing so, or charged for the damage it causes.

Where side-effects are not registered as costs or revenues by the party who causes them, he has no economic incentive to take them into account in making production decisions and his actions will not be consistent with the broader public interest. Generally, too much of the external diseconomies, such as pollution, and too little of the external economies, such as aesthetic amenities, are generated because their benefits and costs are ignored by those who generate them. Such non-marketed side-effects prevent private producers from taking account of the full social benefits and costs of their actions, resulting in inefficient resource allocation.

Perfect Knowledge

In order that the market system can efficiently allocate all resources, consumers must have perfect knowledge about the goods and services available to them, their prices, and the satisfaction to be derived from them. And producers, for their part, must have perfect information about factor markets and the technology of production. In reality, of course, knowledge is always more or less faulty, and ignorance in these matters will lead to faulty allocation of resources.

Even more stringent is the requirement that efficient utilization and production over time calls for perfect knowledge of product and factor markets not only in the present but also in all future periods. That is, there must be no market or technical uncertainty facing producers or consumers.

In forestry, however, uncertainty is a particularly important influence because decisions must be based on long-term plans and projects. Uncertainties about future resource supplies, rates of growth, natural losses, and technological change bear heavily on decision-making. Some of this ignorance can be overcome by increased effort in data collection and research, but these efforts to improve knowledge all involve a cost, and they can rarely eliminate uncertainty altogether.

Equitable Distribution of Income

We have already emphasized that economic efficiency and equity in the distribution of income and wealth are two separate considerations, though changes in one have implications for the other. Equity is a subjective concept, resting on social and political attitudes towards fairness. It cannot be defined by economic analysis, but rather must be resolved collectively through political processes.

Nevertheless, for market processes to allocate resources efficiently, the distribution of income must be optimal in the sense that it is equitable. This is because the distribution of income governs the pattern of demand for goods and services and hence also the efficient allocation of productive resources. The most efficient allocation of resources therefore depends on the distribution of income. So a market economy can respond reliably to social values only if the distribution of income corresponds to what the society considers equitable.

If some group in society—farmers for example, or retired people—receive too small a share of the national income in this sense, then the demand for the goods and services they buy will be less, and fewer of them will be produced than if the group's income were higher. Correspondingly, production of other things will be greater. It follows that if their incomes were raised, resources would be reallocated as a result of the changed pattern of demand. So, if the distribution of income is faulty, resources will be misallocated. Only when income and wealth are equitably distributed will the amount that individual consumers are willing to pay for goods and services provide an accurate measure of the social benefits derived from producing and consuming them.

With an optimal income distribution, the market price that consumers are willing to pay for the marginal product will equal its marginal social benefit, i.e.:

$$VMP = MSB$$

If the distribution of income were not optimal, the willingness of consumers to pay for different products would not correspond to social priorities and so this equality would not hold.

Marginal Conditions for Efficiency: Summary

We can now summarize the conditions for maximum social efficiency in a market economy in a series of equalities:

MSB = VMP = MRP = MFC = VMF = MSC

where MSB = marginal social benefit,
VMP = value of the marginal product,
MRP = marginal revenue product,
MFC = marginal factor cost,
VMF = value of the marginal factor,
MSC = marginal social cost.

All the variables pertain to one unit of a representative factor employed at the margin of production. The first equality on the left-hand side implies an optimal distribution of income. Then the value of the marginal product produced by employing another unit of the factor, as indicated by the amount consumers are willing to pay for it, accurately measures its marginal social benefit (MSB = VMP).

The second equality will hold as long as perfect competition prevails in the product market, so that consumers' value of the marginal product equals the producer's marginal revenue product from employing the additional factor (VMP = MRP). The next equality is the condition for profit maximization, indicating that the producer's additional revenue is equal to the marginal factor cost he incurs to generate it (MRP = MFC). In conditions of a perfectly competitive factor market this marginal factor cost will equal the price that the producer must pay for an additional unit of the factor, or the value of the marginal factor (MFC = VMF). Finally, the value of the marginal factor will be equal to its marginal social cost if all the other equalities hold so that the factor earns its opportunity cost at the margin in all its uses (VMF = MSC).

All of these equalities must hold if resources are to be optimally allocated in a market economy. We noted earlier in this chapter that the ultimate criterion for maximizing social welfare is the equality between marginal social benefit and marginal social cost, the first and last terms in the series, which implies that all factors of production earn their opportunity cost and no reallocation of any factor can increase the social benefit because it would involve a cost equal to the benefit it could generate. Perfect markets, meeting all the conditions outlined above, would ensure that all other terms were made equal.

This formulation of efficiency conditions follows that of the economic theorist Abba Lerner, who attempted more than forty years ago to summarize the rigorous conditions that must be met in a market economy in order for it to succeed in maximizing social welfare. Such an economy could achieve this result only if certain preconditions, noted above, hold as well, including perfect knowledge among producers and consumers, institutional arrangements that give producers complete control over their inputs and outputs, no externalities, convergence of market and social rates of time preference, and an equitable distribution of income. It also rests on certain technical relationships in production, such as the divisibility of inputs and outputs, decreasing returns to scale, and diminishing returns to all factors of production, some of which are examined in the Appendix to this chapter.

The focus on *marginal* conditions is critical. It should be noted that criteria for efficiency say nothing directly about how much, in total, of any productive factor should be devoted to one use rather than another, nor does it hinge on which use generates the most value in total. As in so much of economic analysis, the emphasis is on adjustments *at the margin*. The issue is seldom whether to produce one product rather than other; it is whether to have a little more of one and a little less of the other. Similarly with inputs—the problem is not to find the use to which land or labour or any other input can be most advantageously put in total, but *how much* of it should be allocated to each of its productive uses. Thus when a variety of products and services can be produced from forests, or inputs like land, labour, and capital can be used in a variety of forestry or other productive activities, attention must be focused on the costs and benefits *at the margin* in each form of production.

MARKET FAILURES AND GOVERNMENT INTERVENTION

We have noted various ways in which mixed capitalistic economies fail to meet the stringent conditions necessary to achieve maximum social efficiency. Nevertheless, the theoretical model of a perfect market system provides a concise framework for examining the sources of market failure which provide much of the substance of forest economics dealt with in the remainder of this book.

How well markets perform, how serious their imperfections are, and how much we can benefit from governmental intervention are issues of endless economic and political debate. The answers inevitably differ in different circumstances, and for people having differing political attitudes. This chapter has concentrated on the stringent conditions for perfectly competitive markets and on the major defects of markets affecting forestry, but this is in order to identify the issues, not to belittle the capability of even imperfect markets to allocate resources and advance human welfare.

As already suggested, government intervention is not always aimed at improving efficiency in the allocation of resources; it may be directed toward a better distribution of income, stability, narrow political purposes, or other objectives that may involve some sacrifice in efficiency. Thus governmental actions that result in some loss in efficiency are not necessarily unwanted or contrary to the public interest.

Intervention itself is never costless. It involves not only costs of public administration but also private costs of adjustment and compliance. Because of this, the existence of a market failure is insufficient rationale for intervention, even on efficiency grounds. A gain in efficiency will result only if resulting benefits exceed all the costs of intervening.

It is also important to recognize that, whatever the objective, the interventions of governments are not always well chosen or effective. Governments, like private enterprises, sometimes suffer from ignorance, make mistakes, or find it difficult to adjust to changing circumstances. Thus we can expect to find examples of public policies which do not succeed in advancing their apparent purpose, or do so less effectively than alternative measures available to governments, or cause unintended side-effects.

Nevertheless, western industrial nations have turned increasingly to governments to correct or offset weaknesses in their market economies. They have invoked fiscal measures, regulation of economic activity, public ownership of resources, and sometimes direct governmental enterprise in order to improve economic performance. All these types of intervention are prevalent in forestry.

Markets diverge in so many ways from the conditions necessary to achieve maximum social benefit that we cannot rely solely on markets to determine the allocation of resources. Imperfections in the market processes that guide the development and use of forest resources, particularly, leave plenty of scope for improvement through well-designed governmental policies, institutions, and regulations. Much of the discussion in later chapters deals with the economic implications of governmental interventions affecting the management and use of forest resources.

REVIEW QUESTIONS

1 What is the "opportunity cost" of land and labour used in producing timber? How does competition for these inputs help to ensure that their prices reflect their opportunity costs?
2 Give examples of how land, labour, and capital are substitutable in timber production. Are these inputs subject to the law of diminishing returns?

3 Describe how the substitutability of one input for another must be compared with their relative prices in order to determine least-cost way of achieving a particular level of production.
4 The application of fertilizer to a tract of forest land can increase the yield of timber depending upon how many kilograms of fertilizer is applied per hectare, as follows:

amount of fertilizer kg/ha	increase in timber yield m^3/ha
50	7.5
100	12.5
150	15.0
200	16.0

The cost of fertilizing consists of a fixed cost of $25 per hectare plus $.5 per kilogram of fertilizer used. The timber is valued at $10 per cubic metre (which we will assume is realized in the same year that the fertilizer is applied).

Calculate, and plot on a simple graph, the total revenue and cost of applying these different amounts of fertilizer, and on another graph the corresponding marginal revenue and cost. What is the most efficient level of fertilization? Does this example indicate diminishing returns to fertilization?
5 If fertilizing a forest would cause pollution of waterways, how would the decisions of profit-maximizing forest owners fail to maximize the benefits to society as a whole? What term is used to describe this kind of market failure?
6 In a controversy over the clearing of forest land for agricultural purposes, proponents pointed out that agriculture contributed more to regional income and employment than did forestry. Why is this unconvincing evidence that the change in land use was beneficial?

FURTHER READING

Boyd, Roy G., and William F. Hyde. 1989. *Forestry Sector Intervention: The Impacts of Public Regulation on Social Welfare.* Ames: Iowa State University Press. Chapter 1

Gregory, G. Robinson. 1987. *Resource Economics for Foresters.* New York: John Wiley & Sons. Chapters 5 and 6

Herfindahl, Orris C., and Allen V. Kneese. 1974. *Economic Theory of*

Natural Resources. Columbus, OH: Charles E. Merrill. Chapter 2
Lerner, Abba, P. 1946. *The Economics of Control: Principles of Welfare Economics.* New York: Macmillan. Chapter 6
Mansfield, Edwin. 1982. *Microeconomics: Theory and Applications.* 4th ed. New York: W.W. Norton. Chapters 1, 6-9, and 16. Or, see Chapters 1, 6-8, 12, and 14 in the shorter 4th ed.
Nautiyal, J.C. 1988. *Forest Economics: Principles and Applications.* Toronto: Canadian Scholars' Press. Chapter 3, 5, 6, and 18
Randall, Alan. 1987. *Resource Economics: An Economic Approach to Natural Resource and Environmental Policy.* 2nd ed. New York: John Wiley & Sons. Part 2
Sedjo, Roger A. (ed.). 1983. *Governmental Interventions, Social Needs, and the Management of U.S. Forests.* Washington, D.C.: Resources for the Future. Part 2

APPENDIX TO CHAPTER 2
The Basic Economic Theory of Production

This Appendix summarizes some of the basic concepts of the economic theory of production. It does not cover all the relevant theory, but rather ties together the economic relationships most critical in understanding forest production, which are referred to in subsequent chapters. A thorough discussion of the conventional theory of production can be found in most textbooks on microeconomics.

The following review deals first with technical relationships in production; that is, how technology enables things to be produced by combining inputs in various ways. It then shows how these technical relationships, coupled with information about the costs of the inputs, can be used to identify the most efficient combination of inputs for producing a particular product. Finally, it turns to the efficient allocation of resources across the range of their potential uses.

Production functions. The relationship between inputs and the output in a production process is referred to in economics as a *production function*. In its most general form, the quantity of a product produced per period, Q, depends on the quantities of the inputs, $x_1, \ldots, x_n$, employed in the productive process. This is represented algebraically as:

$$Q = f(x_1, x_2\, x_3 \ldots x_n).$$

Production functions can take complicated mathematical forms to account for the substitutability of one input for another, but here we are concerned only with the general relationship between inputs and outputs and among inputs.

Substitutability of inputs. Nearly any product can be produced by using the inputs in its production function in varying proportions. For example, a particular volume of timber can be produced on a tract of land with a small input of capital equipment and a large quantity of labour, or with less labour and a larger quantity of capital. Considering only these two inputs, the production function for labour, L, and capital, K, used to produce timber, Q, can be written

$$Q = f(L, K)$$

This substitutability among inputs is illustrated graphically in the upper quadrant of Figure 1. The curve ab is an *isoquant*, that is, a curve joining up all the combinations of two inputs that are capable of producing a particular quantity of output. Each point along ab indicates a quantity of labour and capital which can be combined to produce the same quantity of timber on a tract of land. One such combination on the curve is OK capital and ol labour. Points further to the right on the curve indicate less capital and more labour, and vice versa to the left.

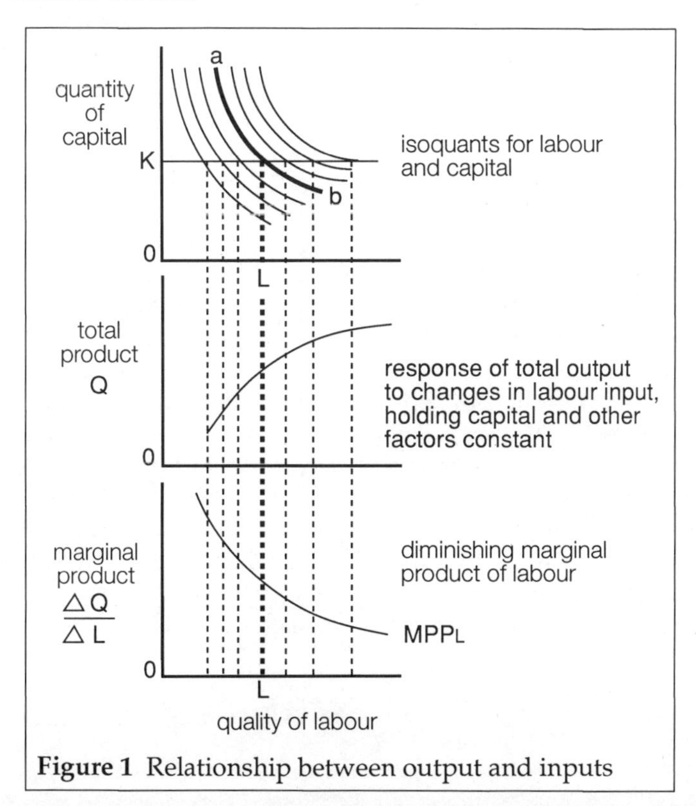

Figure 1 Relationship between output and inputs

The curvature of an isoquant reflects the substitutability of the two inputs or, more precisely, the *marginal rate of technical substitution* of one input for the other. If they are highly substitutable the curves approach straight sloping lines; at the other extreme, if the inputs cannot be substituted for each other at all, the curves become right angles with straight vertical and horizontal sides. In most productive activities, the relevant isoquants fall between these extremes, as shown in the upper quadrant of Figure 1. The advance of technology tends to increase the substitutability of inputs in production processes. For example, developments in technology and specialized equipment have significantly increased opportunities to substitute capital for labour in silviculture.

Typically, the marginal rate of substitution of one input for another is not constant. As more labour is substituted for capital, it will take progressively larger increments of labour to maintain the same level of output, which explains the decreasing slope of the isoquant. This illustrates the *law of diminishing marginal substitutability*, or the general rule that it becomes increasingly difficult to substitute one input for another in producing a given level of output.

The isoquants above ab in Figure 1 represent higher levels of output; those below apply to lower levels of output. As long as the labour and capital are divisible into small units the curves are smooth, and will form a generally symmetrical series of contours, as shown.

To examine the effect on total output of varying one input while holding the other constant, assume that each successive isoquant in the upper quadrant of Figure 1 represents an equal increment of output above the isoquant below it. If the amount of capital is fixed at OK, it can be seen that successively higher isoquants, or levels of output, can be reached only with progressively larger increments of labour. This is reflected in the middle quadrant of Figure l, which shows how total output increases with larger inputs of labour while other factors are fixed. The dashed vertical lines represent equal increments of output (from the top quadrant), so the vertical rise of the total product curve between each is equal. But because progressively larger increments of labour are needed to produce successive equal increments of output, the slope of the curve diminishes to the right.

This illustrates the *law of diminishing returns* or, more specifically in this context, the *law of diminishing marginal product* of one input when others are fixed, illustrated graphically in the lower quadrant of Figure 1. This law states, generally, that if one input in a production process is increased by successive equal increments while other inputs are held constant, output will increase, but the resulting increments of output will become smaller and smaller.

Efficient factor proportions. Isoquants and production functions illustrate the variety of input combinations that can be used to produce a given level of output, but they do not indicate which combinations are economically efficient. This requires information about the cost of the inputs.

For a given cost, it is possible to acquire varying quantities of capital and labour. In Figure 2, the cost of an amount of capital OA would alternatively be sufficient to hire a certain amount of labour, shown as OB. Any point along the line ab represents a combination of capital and labour, such as OK capital and OL labour, that can be obtained for the same total cost. Such a line is referred to as an *isocost curve*.

It follows that the ratio of OA to OB, and hence the slope of the isocost curve, is equal to the ratio of the price of capital to the price of labour. The isocost curve will be a straight line as long as the prices of the inputs are constant and unaffected by how much of each is employed.

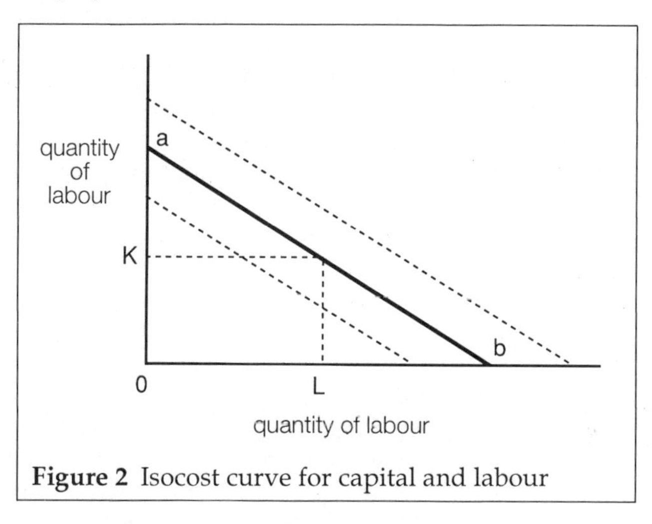

Figure 2 Isocost curve for capital and labour

Isocost curves reflecting higher total costs can be drawn parallel and to the right of AB in Figure 2, and lower total costs to the left. A set of such isocost curves can be represented by the total cost expression

$$TC = P_K \bullet K + P_L \bullet L.$$

Thus the total cost, TC, of a production process depends on the quantities of capital, K, and labour, L, employed and the price, P, of each.

The least-cost combination of inputs can be identified by superimposing isocost curves on isoquants, as shown in Figure 3. For any level of output, such as that represented by the isoquant ab, there is a particular combination of capital and labour that will minimize the total cost of these inputs. This is shown geometrically by the point at which the isoquant is just tangent to an isocost curve. In the figure, this point indicates that the combination of OK capital and OL labour is the most efficient means of producing the level of output represented by ab. The tangency of an isoquant and an isocost curve implies that the substitution ratio between the two inputs, or the ratio of their marginal physical products (MPP_L/MPP_K, reflected in the slope of the isoquant) is just equal to the ratio of their input prices (P_L/P_K, the slope of the isocost curve). That is,

$$\frac{MPP_L}{MPP_K} = \frac{P_L}{P_K}$$

Such a least-cost combination of inputs exists for any possible level of production. A line drawn through these successive optima, shown in Figure 3, is called an *expansion path*. It shows how the most efficient combination of inputs varies over different levels of output.

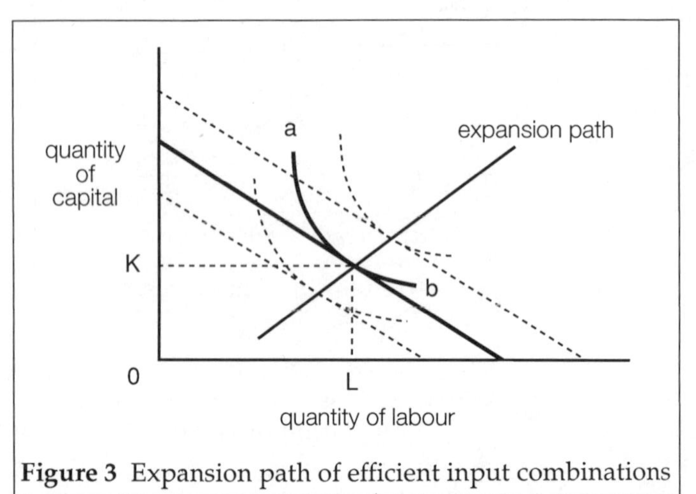

Figure 3 Expansion path of efficient input combinations

Returns to scale. In making production decisions producers must pay careful attention to how their costs are affected by the level of output or scale of operations. In some cases, output is proportional to the inputs used, which is referred to as *constant returns to scale*.

Successive equal increments of inputs will generate equal increments of output, so the isoquants will be equally spaced.

In other cases, the additional output generated by an increment of inputs decreases as the scale of operations is expanded, referred to as *decreasing returns to scale*. In these circumstances, where output increases less than proportionally to inputs, the isoquants for successive equal increments of output become more widely spaced at higher levels of output. The opposite case is that of *increasing returns to scale*. The expansion path will form a straight line only under constant returns to scale.

The precise relationship between outputs and inputs is a critical influence on the producer's choice about his level of production because it determines his marginal cost; that is, the additional inputs required to increase output by a small amount multiplied by their input prices. In many production processes the returns to scale vary, sometimes increasing over a range of output, causing marginal cost to decline, and decreasing over another range, causing marginal cost to rise. Producers have strong incentives to expand their operations until they exhaust increasing returns to scale. Beyond this point, decreasing returns and rising marginal cost limit the scale of efficient operations and, as we shall see below, play a crucial role in maintaining competition among producers. We will ignore, here, the special cases of constant and increasing returns to scale which imply no limit to the efficient scale of operations, and invite all production to fall to a single producer.

Profit maximization. The points of tangency between isoquants and isocost lines illustrate how producers choose the cost-minimizing combination of inputs in producing any quantity of output. It remains to examine how they choose their level of output.

In order to maximize their profits (π) or net returns, producers must find the level of production that will generate the greatest surplus of total revenue ($TR = PQ$) over total costs ($TC = P_L L + P_K K$). This implies minimizing input costs, as explained above, for a level of output selected with reference to marginal cost and marginal revenue.

As we have seen, the *marginal cost* of production is the additional cost of inputs required to produce a unit of output. The *marginal revenue* from a unit of output sold in a perfectly competitive market is reflected in its price. To maximize profits, producers must expand their output as long as their marginal revenue exceeds their marginal cost. However, as long as the marginal cost rises with output, there will be a point beyond which marginal cost exceeds marginal revenue, implying a net loss at the margin of production. Therefore, the profit-maximizing level of output is that at which marginal cost, MC, just equals marginal revenue, MR, i.e.,

$$MC = MR.$$

This profit-maximizing behaviour explains why producers are willing to produce and sell more at higher prices, indicating upward sloping market supply curves for products. When the price of a product increases, a firm's marginal revenue will be raised above its marginal cost, and in order to maximize profits under these changed conditions the firm must expand output, moving upward along its marginal cost curve until equality between the two marginal variables is restored. Thus the firm's supply curve slopes upward, following its marginal cost curve, and the market supply curve is the sum of the supply curves of all the producing firms. So, as a general rule, when the prices of forest products rise, more of them are produced and offered for sale.

These relationships also explain why producers will demand more inputs when input prices are lower. When input prices fall, a firm's marginal cost is lowered. To maximize profits the firm must restore equality between its marginal cost and marginal revenue by expanding production, which will require more inputs. Thus the demand curves for inputs slope downward in the usual fashion, consistent with the observation that producers employ more labour, capital, and other inputs in production when their prices are lower.

The main part of this chapter explains how profit-maximizing producers will employ more of any factor of production as long as its marginal revenue product exceeds its cost or price, because this will enhance profits. In a perfectly competitive economy, all producers using a particular factor will employ it up to the point at which its marginal revenue product equals its price, so that the value it generates at the margin is the same in all its uses. Through this process all factors of production in a competitive economy become allocated so that they generate a value at least equal to their opportunity cost in every use, and no reallocation could increase the aggregate value of production. This implies that the total value of production is maximized and factors of production are efficiently allocated.

CHAPTER THREE

Timber Supply, Demand, and Pricing

Out in the Forest . . .

The market for logs produced on Sundry Island ultimately depends on the supply and demand for consumer goods made of wood, such as housing, toilet tissue, and hundreds of other products in distant, often foreign, markets. Producers of all these products, including logs, find that the quantity they can sell depends on the price—more can be sold the lower the price.

The president of Peavey Forest Products Limited closely watches log price fluctuations and trends. When prices rise, David Cameron's concern for profits compels him to immediately consider expanding production, employing extra men and working his equipment longer hours; he follows the reverse strategy when prices fall. If a new price is sustained for long enough it is to his advantage to adjust even more, by expanding or reducing his equipment, roads, and other capital facilities as well as his variable inputs.

Forecasts of long-term trends in prices of forest products are provided to the company by its industry association. Recent analyses have convinced Cameron that, apart from market swings, the long-term trend in prices of logs and timber is upward. The outlook for higher timber values has meant that Ian Olson can justify higher spending on silviculture and forest protection. This will add to the company's future growing stock, but only over many years. So the ways in which the company can adjust to a price change depends on how much time it has to respond.

In response to the higher expected timber values Olson has also recalculated the company's inventory of merchantable timber, to include some stands that were previously excluded because the recoverable timber would have been worth less than the cost of logging it. In this way, as well, higher prices result in expansion of the company's production capacity.

One of Olson's main responsibilities is to keep track of the company's forest inventory as it changes over time, increasing through growth and any improvements in recoverability, and decreasing through natural losses, harvesting, and any deletions from the land base available for timber production.

All of these changes are influenced by economic forces which are hard to predict, complicating his task of planning the company's long-term timber supply.

Of the many products and services produced from forests, timber is the dominant industrial product. Accordingly, much forest management effort is aimed at producing timber for industrial raw material. This chapter examines the economic forces that govern the value of timber.

The simple answer to the question of what determines the value of timber is the same as it is for other goods and services: supply and demand. The market price of timber will adjust to the level at which supply meets demand. But the concepts of supply and demand are not as straightforward for timber as they are for many other products, and so call for some special consideration.

THE DEMAND FOR TIMBER

The demand for timber is a *derived* demand. That is, it is derived from the market demand for final goods produced from wood. Final products are those wanted by consumers; those that depend most heavily on wood are familiar: housing, newspapers, toilet tissue, wrapping paper, furniture, and so on. The consumer demand for these products creates a demand for the materials needed to produce them; and in this sense the demand for timber standing in the forest is derived from the demand for consumer products that can be made from wood.

The derived demand for timber is illustrated in Figure 4. The upper quadrant shows the demand for newspapers, a final product which uses timber as a raw material. The supply curve represents the supply cost of all inputs in the production of newspapers except raw timber, including the cost of processing wood into newsprint. To the left of the intersection of the two curves, the excess of the amount that consumers are willing to pay for newspapers over the cost of all other factors represents the maximum amount that producers would be willing to pay for timber. Thus the derived demand for timber, shown in the lower quadrant, reflects the difference between the two curves in the upper quadrant.

Similarly, a demand for timber is derived from all other products that use it as a raw material—housing, packaging, and so on. The sum of all these derived demands is the market demand for timber.

Figure 4 makes it clear that the derived demand for timber, like the demand for final products, follows the "law of demand"—more will be purchased per period at lower prices. Hence demand curves for timber slope downward in the usual fashion, as illustrated in Figure 5.

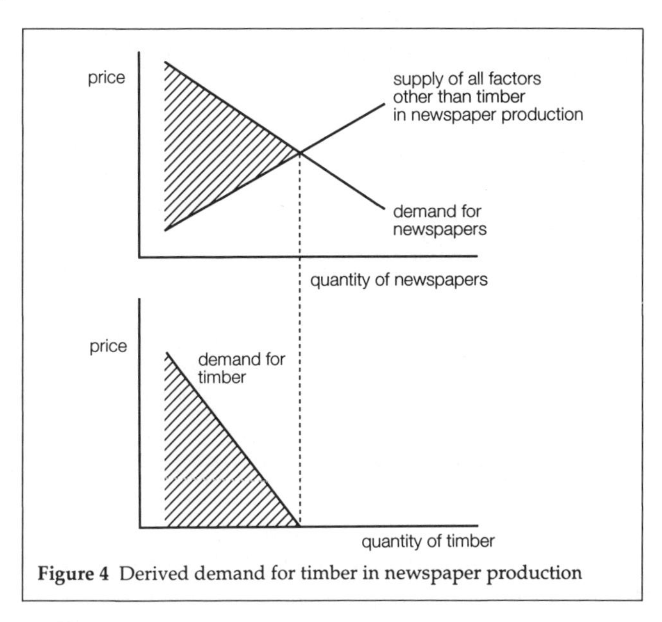

Figure 4 Derived demand for timber in newspaper production

The supply of timber, discussed below, slopes upward to the right in the usual fashion, and its intersection with market demand yields a market price. This price, included with the cost of all the other factors in newspaper production, will yield an equilibrium price for newspapers to the left of the intersection of the two curves in the upper quadrant of Figure 4.

The derived demand for *primary* resources like timber is usually generated through several steps, involving the production of *intermediate* products needed to produce *final* goods and services. For example, the demand for housing generates a demand for plywood, which in turn generates a demand for logs suitable for making veneer, and the demand for veneer logs creates a demand for suitable timber.

For each intermediate product—plywood veneer and logs in this example—there is usually a distinct industry and market with its own supply and demand relationships. However, in some cases the wood industries are so integrated that there are no separate markets for intermediate products. Some firms that produce plywood, for example, manufacture it directly from timber they harvest themselves, thus eliminating intermediate markets for logs and veneer.

Any change in the demand for final products made of wood will change the derived demand for timber. However, it is important to note that this relationship is not rigid; a change in the demand for a particular final wood product cannot be expected to generate a proportional shift in the demand for standing timber, for several reasons. First, a stand of timber is typically best utilized in producing a *mixture* of products, such as plywood, lumber, and pulp, in proportions that can be altered in response to changes in the relative prices of the products. For example, if the demand for housing doubled it does not follow that the demand for timber would double, because only a portion of timber is manufactured into construction materials and if the demand for other products did not change the increase in demand for timber would be less than proportional. In addition, producers could be expected to shift some of the timber previously more profitably utilized in other products into increased production of construction materials. Finally, a doubling of the demand for housing would increase the demand for all of the inputs in housing construction and drive up their prices, but not proportionally. The result would be substitutions and changed proportions of labour, capital, and various kinds of materials used in construction, so the impact on the markets for inputs would vary.

In the long term, the relationship between the demand for timber and the demand for final wood products will change as a result of advances in technology. Consider, for example, the linkages between the demand for timber and the demand for ships, small boats, multiple-unit housing, offices, furniture, sports equipment such as skis and fishing rods, and heating fuel; all these products were formerly made largely from wood, but technological changes have almost eliminated wood as a major component in them. On the other hand, there are the familiar new products made of wood such as fabrics, paper milk containers, and drinking cups, and various chemicals. Moreover, the kinds of raw materials needed to manufacture wood products have changed significantly, often toward the utilization of previously waste material, such as sawmill by-products and low-grade species now used for making pulp and hardboard. Over the long periods involved in producing timber and planning long-term timber supply, discussed in Chapter 8, such technological changes can dramatically alter the relationship between demands for final products and the demand for timber.

The responsiveness of the quantity demanded to a change in price is called the *elasticity of demand*. It is measured at any point along a demand curve as an elasticity coefficient, Ed, calculated as the percentage change in quantity demanded divided by the corresponding percentage change in price. If a 5 per cent increase in price results in a 5 per cent decrease in quantity demanded the elasticity coefficient has a value of one (Ed = .05 ÷ .05 = 1). When the coefficient is less than one, demand is said to be inelastic; when it is greater than one, elastic. At the extremes, a coefficient of zero implies no demand response to a price change, while an infinite value implies that a small increase in price would eliminate all demand for the product. The elasticity of demand almost always lies between these extremes, and differs at different points along a demand curve. Corresponding measurements can be made of the *elasticity of supply.*

The elasticity of demand is an important indicator of how the total revenue of producers will change as a result of a change in price. If the elasticity coefficient has a value of one, the total revenue—price multiplied by the quantity demanded—will be the same before and after the change. If its value is less than one, an increase in price will increase revenue, and if it exceeds one, a price increase will cause revenues to decline.

To summarize, for any standing timber of industrial value there is a market, and the demand side of this market can be depicted by a demand curve of the customary downward-sloping type, implying that more will be purchased per period at lower prices. This demand for timber is generated by—hence *derived* from—the demand for final products that use wood in their production. The relationship between the demand for final products and the demand for timber is determined by the relationships described in the Appendix to Chapter 2; the production functions relating inputs with outputs, the substitutability of inputs in production processes, and their relative costs.

TIMBER SUPPLY

The meaning of "timber supply" deserves attention, because this term is often used loosely and confusingly to refer to the forest inventory or to the outlook for future harvests. Here, we use it in the conventional economic sense to refer to the quantity of timber that will be supplied in a market, per period, at various prices. Normally, the higher the price the greater the quantity supplied, of timber as with other products. But in the case of timber it is especially important to be clear about the cause and effect and the period referred to.

A *short-run* supply curve, such as that illustrated in Figure 5, indicates how the quantity of a product supplied in a market will vary in response to price within a period too short to alter the physical capital used in production, such as logging facilities, pulp mills, and sawmills.

In the short run, producers can alter output only by changing their variable inputs such as labour, fuel, and raw materials, using their existing capital facilities more or less intensively. To maximize their profits without being able to alter their fixed costs, producers will choose the level of production at which their marginal revenue is just balanced by their marginal variable costs, or short-run marginal costs.

The short-run market supply curve is thus the sum of the short-run marginal cost curves of all the producers serving the market. Because producers cannot adjust their physical capital within this short period, short-run supply curves are relatively inelastic with respect to price.

The dividing line between the short term and the *long term* depends on how long it takes to change the capital required in producing the product. This will inevitably vary depending on the product and the technical requirements for producing it. The long run might be less than a year for the logging industry compared to several years for the pulp and paper sector because the capital equipment used in producing logs is relatively mobile and can be obtained and put in place quicker than that required for manufacturing pulp and paper. But whatever the period required to change the capital involved, the industry will be able to respond more flexibly to a change in price in the long run than in the short run, and hence the long-run supply curve is always more price-elastic, as Figure 5 indicates.

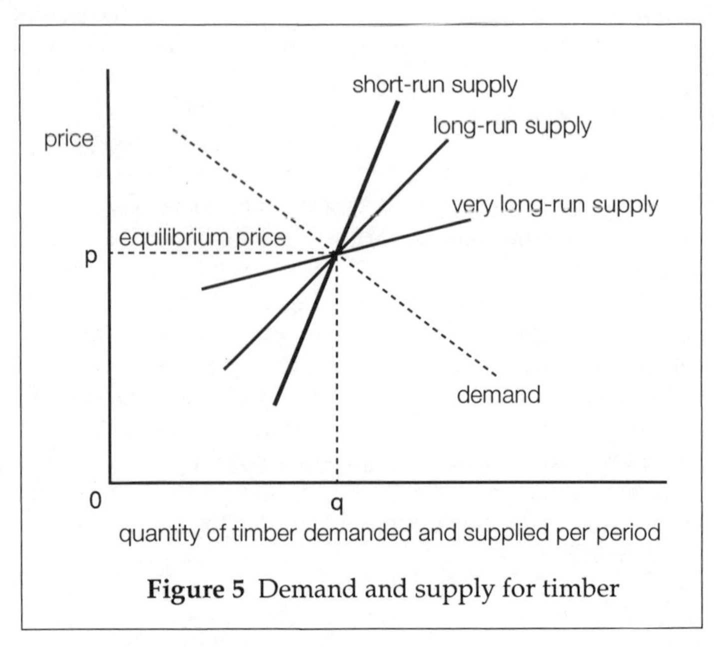

Figure 5 Demand and supply for timber

The distinction between short-term and long-term supply response is important in analysing the impact of price changes in any industry that employs physical capital in production. However, in timber production, there is an even longer run to consider because it involves capital not only in the usual form of plant and machinery but also in the form of forests. Like other capital, the stock of forest capital can be run down through harvesting or built up through investment. The difference is that the time required for forest management is usually much longer than the time it takes to build a factory.

An increase in the price of timber, if it is expected to be permanent, will attract more land into industrial forestry and induce more intensive management. After many years this will increase the market supply of timber. But for the timber production sector the term of sufficient duration to enable adjustment of forest inventories to changes in timber prices must be at least a complete crop cycle. In this *very long term*, the timber supply curve is even more elastic than in the customary long term.

To summarize the preceding paragraphs, three types of supply response to changes in the price of timber can be distinguished on the basis of the adjustment period to be considered and therefore the range of inputs that producers can manipulate. In the *short term* they cannot change any capital, so they are constrained to alter only variable inputs, using existing capital more or less intensively. Over the conventional *long term*, producers can adjust their production capacities so their response to price changes is more flexible. Over the *very long term*, long enough to grow more timber, producers can adjust all their inputs—variable inputs, physical plant and machinery, and forest capital—and so their supply response to a given price increase is even greater. Figure 5 illustrates each of these three supply responses.

Conventional supply curves of the kind illustrated in Figure 5 trace points of equilibrium, that is, the quantity of the product that producers will seek to supply at various prices within the relevant period. Each point on the curve indicates the level toward which market supply will tend, but markets are more often in the process of adjustment toward equilibrium than at a stable equilibrium. The fluctuations we observe in production reflect the ongoing efforts of producers to adjust to changing market opportunities. Their responses are constrained, as we have seen, by the period of time they have to adjust.

PRICE EXPECTATIONS AND SUPPLY RESPONSES

In most types of production, producers can increase output in response to a rise in their product's current price without having to worry about sacrificing future output. But in forestry any timber harvested today will not be available in the future. The optimal pattern of resource use over time requires that the values that can be realized by harvesting today be compared with the value of the same resources if they were harvested sometime in the future.

Chapter 6 examines methods of comparing values over time; here it is sufficient to note that the value generated from the resource will be increased by harvesting more in the present as long its net value, at the margin, exceeds the value it would generate if harvested, instead, in a future period. However, the future value must be discounted at the owner's interest rate to indicate its equivalent present value. For example, if the owner's interest rate is 5 per cent, timber harvested next year yielding $100 per cubic metre has a present value of $95 per cubic metre. This is the *user cost* of harvesting today, the sacrifice in future returns resulting from harvesting in the present. The owner should postpone harvesting as long as his net return from harvesting in the present, at the margin, falls short of this amount. By thus balancing present and future returns at the margin, he can maximize the value realized from the resource over time.

The response of producers to a change in price may not always follow the orderly pattern implied by a smooth supply curve because much will depend on their expectations about how long the new price will prevail. For example, a temporary increase in the price of timber will make it attractive to harvest more today rather than leave it for the future. As a result, current production will increase, and, by reducing the harvestable timber available, cause a decline in production in future periods. However, if the new price is expected to prevail, future harvests will be more profitable also, raising the user cost of harvesting today and reducing the incentive to shift production toward the present. A higher price that is expected to be permanent will make it profitable to expend more on silviculture and forest management, to harvest stands that were previously beyond the margin of economic recoverability, and to utilize more of the trees harvested. Thus, a price increase that is expected to be permanent will generate a variety of responses, but the net effect will usually be increased timber production in all periods.

The interaction of these supply and demand relationships is sum-

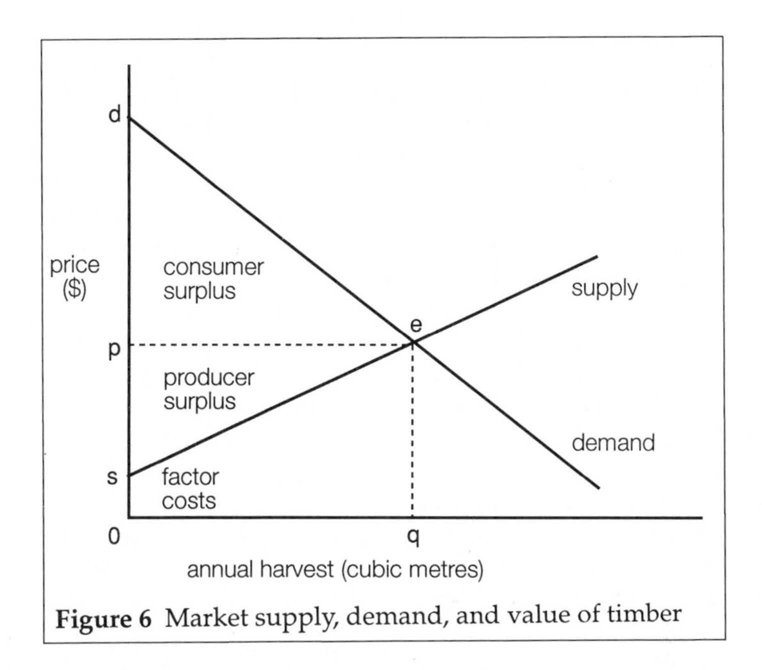

Figure 6 Market supply, demand, and value of timber

marized in Figure 6. In any market for timber, supply and demand determine an equilibrium price, op, and a corresponding equilibrium annual rate of production, oq. If the demand increased, shifting the demand curve upward, both the price and the quantity produced would increase. If production costs fell, shifting the supply curve downward, the quantity produced would increase but the price would fall. In both cases the results would be reversed if the change were in the opposite direction.

PRICE, COST, AND NET VALUE

The total value of the timber produced is the full amount that the demanders would be willing to pay for it. This is reflected in the total area under the demand curve to the left of the point e in Figure 6, which is the area odeq. Subsequent chapters discuss the various components of this value. As long as all production is sold at the equilibrium price, the total payments to suppliers is that price multiplied by the quantity sold, represented by the rectangle opeq. The supply schedule indicates that the cost of the factors of production the suppliers need to produce this timber is represented by the area oseq. These *factor costs* account for only part of the suppliers' total receipts, however. The remainder, represented by the triangle spe, is *producer's surplus*. In timber production, as we shall see later, this surplus usually accrues as rent to the owner of the land. The remaining area between the price line and the demand curve to the left of point e is *consumer surplus*, or the amount that demanders would be willing to pay in excess of the market price.

In terms of Figure 6, the *net* value of forest production is the gross value minus the supply costs, or the sum of producer and consumer surpluses. However, if the producers surplus accrues as payments of rent to owners of land which has other possible uses, the net gain from forest production must be reduced by the rent the land could earn in its next highest use, its opportunity cost, as discussed in Chapter 4.

It is important to take account of producer and consumer surpluses when the market's entire production, or a significant portion of it, is at stake. However, forest managers are more often required to make decisions that would increase or decrease the supply of timber or some other forest product by a small fraction of the market's total supply. In these cases, where the task is to assess a marginal change that will not significantly affect the market price, there will be no measurable change in these surpluses and they can be ignored.

LONG-TERM WOOD SUPPLY PROJECTIONS

In much of the forestry literature, the long-term timber supply refers to the quantity of wood that will become available over time, usually over many decades. A supply projection of this type is depicted in Figure 7; it shows the volume of timber expected to be supplied per period—usually per year—over future years. Such projections may relate to the supply of timber in a region, the supply in a particular market, or the production from a particular forest.

In such projections the trajectory of timber production over future decades is estimated on the basis of the age composition, rate of growth, and other characteristics of the forest inventory, often in the context of some harvest regulation policy designed to sustain yields.

It is important to recognize the difference between such supply projections and the economic concept of supply reflected in the supply curves discussed above. These long-term timber supply projections depict the quantity supplied as a function not of price but of time. The economic assumptions embodied in them are often unclear, so the results must be interpreted cautiously.

Market forces will generate a supply of timber over time according to such a projection only if every point along the trajectory repre-

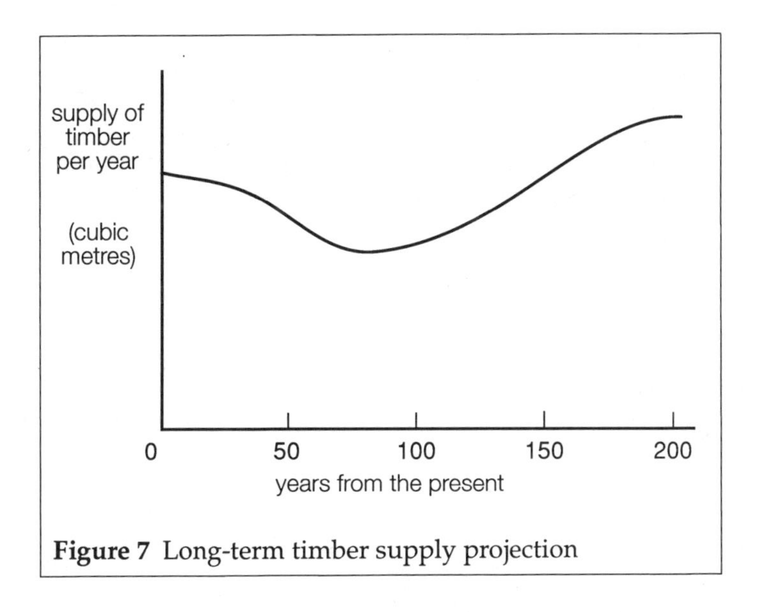

Figure 7 Long-term timber supply projection

sents an equilibrium of supply and demand at the corresponding time. If the projection is not based on these market forces, or ignores likely trends in prices, costs, and technology that influence timber producers, the projected results can be realized only through governmental control. Indeed, projections of this kind are often based on artificial yield controls or policies, discussed in Chapter 8, and in that context they may represent only some regulated upper limit on timber production.

TIMBER SUPPLY AND VALUE OF THE FOREST INVENTORY

Typically, some timber in a market supply region cannot be profitably recovered and used because it would cost more to harvest than the recovered wood is worth. This is usually the case in frontier regions with extensive natural forests. An important determinant of the long-term timber supply is the proportion of the total forest inventory that is, or can be expected to become, worth harvesting. Supply projections of the kind referred to above must therefore be based on estimates, or assumptions, about the economically recoverable portion of the total inventory.

The forest inventory in a market supply region consists of stands of timber that vary in terms of their economic value. Differences in the technical characteristics of stands such as species composition, size, and other qualities yield varying market prices on the one hand, and their varying distance from markets, terrain conditions, and so on cause differing production costs on the other hand.

Thus, the timber in each stand has a unique *net value* or *stumpage value* per cubic metre, measured by the difference between the value of the timber that can be recovered from the stand and the cost of harvesting it.

A forest usually consists of many stands, covering a spectrum of net values. They can be arranged according to their net value to produce a curve like that in Figure 8. Each point on this curve indicates the quantity of timber, measured on the horizontal axis, recoverable from stands having a net value per cubic metre equal to or greater than the corresponding point on the vertical axis. The curve must slope downward, but its curvature and shape may take any irregular form, reflecting the structure and value of the forest inventory.

Figure 8 thus portrays a forest inventory in terms of its economic recoverability. Ranking the stands from the left in order of their net value identifies the total volume that has a net value greater than zero. Timber having a net value of zero is at the *extensive margin* of recovery. It is marginal insofar as the recoverable values will just cover the recovery costs. The portion of the inventory having a negative net value is outside the extensive margin of recoverability. The portion within the extensive margin, having a positive net value, comprises the economically recoverable inventory, as indicated in Figure 8.

The economically recoverable inventory is the economic stock of timber from which producers can draw. It may be substantially less than the total physical inventory that includes the uneconomic or sub-marginal portion.

Note that the extensive margin is determined by the balance between values and costs, so the economically recoverable inventory is sensitive to any changes in these economic variables. This is particularly important in making projections of timber supplies over long periods; then we need to estimate not only how much timber has a positive net value today, but also how the extensive margin is likely to shift over time as a result of changes in prices, costs, and technology. Over the long periods common in forest yield planning, such changes can be substantial, putting a heavy onus on assumptions about future trends in costs and prices.

The shaded area in Figure 8 represents the total net value of the entire forest inventory, of which only a small portion is usually harvested in any year. Over time, the inventory is diminished

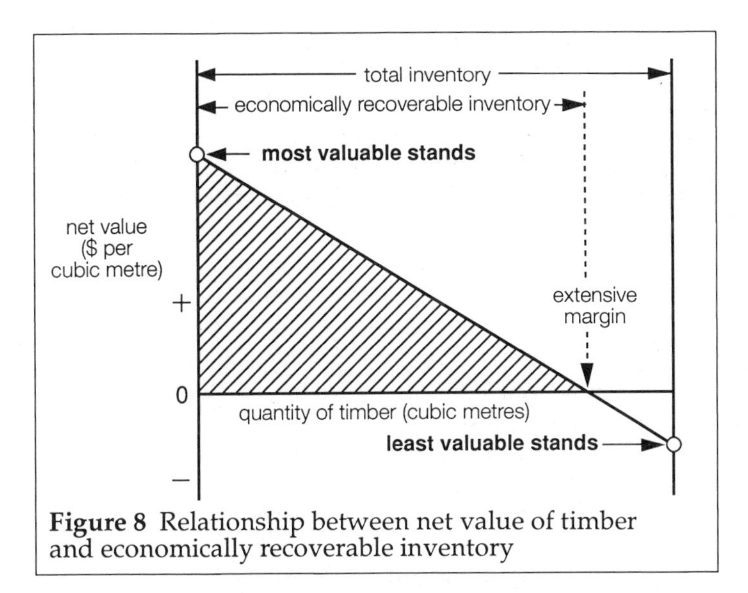

Figure 8 Relationship between net value of timber and economically recoverable inventory

by harvesting and natural losses, shifting the curve in Figure 8 to the left, and increased by growth, shifting the curve to the right. Thus the net effect on the harvestable inventory will depend on the balance between these reductions and additions, both of which are influenced by forest management decisions.

TIMBER SUPPLY AND THE ECONOMIC INVENTORY

The foregoing explains how the merchantable inventory of timber available to a particular timber market at any moment is constrained by its cost of recovery and its recovered value, or price. It remains to examine the most economically efficient pattern of harvesting from the merchantable inventory.

To define the most efficient pattern of harvesting over time we must consider three related issues: the *sequence* in which timber will be removed from the economic profile of the inventory; the *level* at which harvesting will initially take place; and the *change* in the level of harvesting over time. The objective, we assume, is to generate the highest possible return from the inventory which (as explained in detail in Chapter 6) requires maximizing the present worth of harvests.

Initially, let us assume that the merchantable stock of timber is fixed (that is, it does not grow) and can be depicted in terms of its range of net value as illustrated in Figure 8. We will assume also that costs and prices are not expected to change over future time.

The first question is the *sequence* in which timber of varying value should be harvested. As will be shown in Chapter 6, the goal of maximizing the total value of a future stream of harvests requires that the greatest net values be discounted least. This means that the timber with the highest net value should be harvested first, depicted graphically by removing successive slices from the left-hand portion of the inventory in Figure 8 until the extensive margin is reached.

As harvesting progresses in this way, the net value of the timber removed declines because it brings lower prices, it costs more to harvest, or both. In developing forest regions, especially, harvesting costs are likely to rise as operations advance into more remote areas. Progressively higher costs will shift the short-run supply curve upward in successive production periods.

The optimal *level* of harvests is determined by comparing present and future net returns, as suggested earlier. It is the level at which the marginal net revenue (the difference between the price received for the timber and its marginal recovery cost) is just equal to the marginal user cost of the harvested timber (the present worth of the net revenue that could be realized by harvesting the timber in a future period). In a perfectly competitive market, this level is achieved by expanding production until the short-run marginal cost plus the marginal user cost of production rises to the price of timber.

The final issue is that of *change* in the level of harvesting over time. Under our assumptions of a fixed resource stock and constant prices and costs, present worth maximization calls for a declining rate of harvest for two reasons. First, production must be tilted toward the present in order to equate marginal net revenue with marginal user in all periods. Harvests in the more distant future must, because of the force of discounting, generate a higher net return at the margin in order to equate with present marginal net revenues, implying a lower rate of future production. (Note that this would apply even if the entire merchantable inventory was of uniform value.) Second, whenever depletion involves a progression into timber that is more costly to harvest, the short-run cost curve will shift upward, resulting in reduced market supply at any given price.

LONG-RUN EFFECTS OF GROWTH AND DEPLETION

Most forests cannot be realistically characterized as consisting of a fixed merchantable stock of timber, as assumed above. Over time, the inventory is reduced not only by harvesting but also by natural and often unpredictable losses. It also increases with forest growth, which can be enhanced by silvicultural techniques. Changes in prices and costs of timber and in other forest values will shift the extensive margins of production and lead to additions or withdrawals of forest land from timber production, as explored in more detail in Chapter 5.

As a result of all these influences, the curve of net value in Figure 8 will shift and change over time. The curve represents only a snapshot of the forest at a particular moment.

Moreover, when account is taken of growth, it may not always be most efficient to harvest the most valuable timber first. For example, the timber having the highest net value may also be growing rapidly in value, so its high user cost will mean that its harvest should be postponed. As we shall see in Chapter 7, the optimum time to harvest a stand depends not only on its current value, but also on its rate of growth in value and the productivity of the forest site it occupies.

When all these variables associated with actual forests are accounted for, the problem of determining the optimum pattern of harvesting over time is considerably complicated. Recently, modern computer technology has facilitated development of sophisticated dynamic models to enable detailed projections and assessments of forest growth and yield over time. Combined with economic information or estimates relating to costs and prices, these models can identify the pattern of harvesting that will best meet a specified objective. We return to the issue of harvest scheduling in Chapter 8.

PRICE DISTORTIONS

In any timber market, the interaction of supply and demand generates an equilibrium price. But both supply and demand change more or less continuously, causing the shifts and trends in prices that we observe in markets for timber. Suppliers and demanders are constantly adjusting, with lagged responses, to these changes. As a result, the equilibrium or market-clearing price of timber in a particular market at a particular moment is not necessarily the actual price but rather the price toward which the market price is adjusting.

Moreover, timber markets are seldom perfectly competitive, and various imperfections impede the interplay of supply and demand. Monopolies and oligopolies often prevail in local timber markets, distorting costs and prices, artificially restricting timber supplies, and maintaining prices above marginal revenue and cost. Geographic monopsonies are also common, where suppliers face only one buyer and both price and supply are depressed. And wherever firms manufacture into higher products all the timber they harvest, there is no market at all for unmanufactured timber.

Governments also influence market supply, demand, and prices through taxes, royalties, and other fiscal or regulatory policies. Sometimes these interventions are aimed at correcting market failures or redistributing income. Moreover, governmental agencies are often deeply involved in managing forest inventories directly, especially on public lands. They sometimes require that the timber inventory be harvested in a sequence governed by the age of its constituent stands. Examples are the "oldest first" and "old growth first" rules applied to some public forests. More common are sustained yield policies that prescribe more-or-less constant rates of harvesting, examined in Chapter 8. Such harvesting regimes deviate from the economically most efficient pattern because they take no direct account of interest, costs, and prices.

Whenever obstacles to competition or other market imperfections exist, the market price of timber is likely to differ from its social value, or marginal social benefit, as defined in Chapter 2. Moreover, the prices actually paid for timber may differ from those that would result from competitive market forces as a result of the policies of forest landowners. Public timber, which accounts for a major share of the total supply in many regions of North America, is often made available to private users through licensing arrangements at prices that are not determined by market competition. The payments take a variety of forms, including rentals, licence fees, royalties, special taxes, stumpage rates, and other levies. These matters, and the distinction between the economic value of timber and the price actually paid for it, are examined in Chapter 10.

REVIEW QUESTIONS

1 In what sense is the demand for timber a "derived demand"? Describe the connection, if any, between the demand for newspapers in New York and the demand for timber in Quebec.
2 Why does the full market supply response to a change in demand take longer for timber than for most other products?
3 Use a diagram of supply and demand curves to illustrate how improvements in the manufacture of artificial Christmas trees will affect the demand, price, and sales of natural Christmas trees.
4 The accompanying diagram illustrates how total logging costs and total revenue from the harvest increase with the amount of timber recovered from a hectare of forest, assuming that the harvest always consists of the most valuable timber. Identify in this diagram (a) the fixed costs of logging, and (b) the break-even

levels of utilization. Draw the corresponding marginal cost and revenue curves to show the most profitable level of utilization, or the intensive margin of recovery. How would an increase in the price of timber affect this intensive margin?

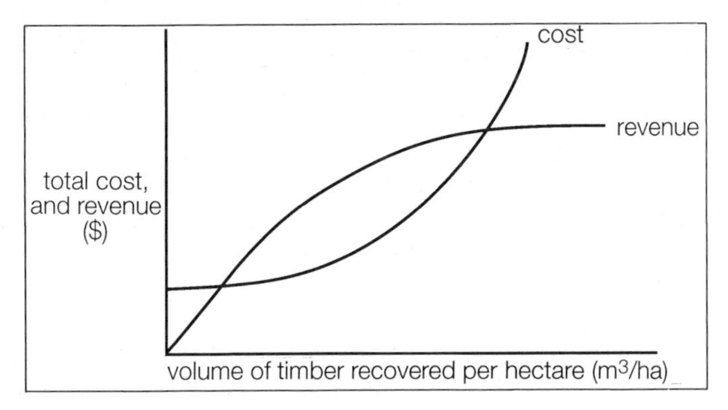

5 What is the difference between a long-term projection of timber production for a region and a conventional market supply curve for timber?
6 What is the "extensive margin" of timber production? Give examples of changes in prices and costs that would shift the extensive margin.

FURTHER READING

Duerr, William A. 1960. *Fundamentals of Forestry Economics*. New York: McGraw-Hill. Part 3

—. 1988. *Forestry Economics as Problem Solving.* [Blacksburg, VA]: By the author. Parts 2, 5

Gregory, G. Robinson. 1987. *Resource Economics for Foresters.* New York: John Wiley & Sons. Chapters 9 and 10

Jackson, David H. 1980. *The Microeconomics of the Timber Industry.* Boulder, CO: Westview Press. Chapter 3

Mansfield, Edwin. 1982. *Microeconomics: Theory and Applications.* 4th ed. New York: W.W. Norton. Chapter 2 (or the shorter 4th ed., Chapter 2)

Nautiyal, J.C. 1988. *Forest Economics: Principles and Applications.* Toronto: Canadian Scholars' Press. Chapters 4 and 5

Vaux, Henry J. 1973. How much land do we need for timber growing? *Journal of Forestry* 71(7):399-403

CHAPTER FOUR

Unpriced Forest Values

Out in the Forest . . .

Sawlogs and pulpwood are the only commercial products produced by Peavey Forest Products Limited, but the company's forests and surrounding forestland provide a variety of other benefits that are not bought and sold in markets. Like the nearby public park, the company's forestland has a significant recreational value, enjoyed without charge by hikers in the summer, hunters in the fall, and cross-country skiers in the winter. Others appreciate the unique example of virgin forest protected in the park. And both the park and the company's forest enhance the value of adjacent property to the benefit of other landowners on Sundry Island.

These non-commercial forest benefits are substantial, and increasing in value, but the absence of markets and prices for them makes it difficult to estimate how much they are worth to the beneficiaries and to the community at large. This is particularly problematical for the Parks Service in assessing how much it should invest in recreational and other improvements to the park, and for the Forest Service in designing regulations to protect environmental values on private land.

In the past, officials of public agencies like the Parks Service and Forest Service have had to depend on their own judgment about these non-timber values, and their regulatory decisions have often been controversial. To provide more guidance, the Forest Service recently has sponsored a series of surveys and studies to estimate how much the outdoor recreation opportunities are worth to those who benefit from them, or how much users would be prepared to pay for them if they were priced. The results are not precise, but they provide valuable economic guidance for planning resource development.

The preceding chapter considered commercial timber as the product of forest management. However, as noted, there is a wide variety of other products and services produced from forests, ranging from livestock forage and water to recreational, aesthetic, and environmental benefits. The importance of these non-timber benefits varies widely; in some forests they are insignificant, while in other forests or parts of forests one or more of them constitutes the dominant value. Almost everywhere the demand for them is growing, their value is accordingly rising, and they are becoming increasingly influential in forest management.

To provide appropriately for these other forest products and services, forest managers and planners must take account of their values and costs of production. Frequently, more of one can be produced only with some sacrifice of another, or of commercial timber production, implying a need for careful compromises in order to maximize the total value generated by a forest. This problem of trade-offs and the economics of multiple use is the subject of the following chapter. But before we turn to the question of integrating the production of various goods and services we must deal with the means of evaluating each, in order to identify efficient levels of production and investment in them. This requires assessments of their value and production costs, which is the subject of this chapter.

UNPRICED VALUES: A PROBLEM OF MEASUREMENT

Some of these other forest products and services are priced and marketed much like industrial timber. For example, livestock forage is sometimes sold to ranchers by means of grazing rights issued at prices determined by competition. In such cases there is no special problem in assessing the value of the benefits other than accounting for market imperfections that may cause prices and social values to diverge.

Other forest values are made available to users without charge. These non-marketed benefits include services, like outdoor recreation and amenity, and products such as game, wild berries, and fuel wood. The range of forest products and services available to people who do not pay for them, in contrast to those that are priced and sold, varies according to the patterns of property rights and governmental policies in different jurisdictions and regions. But in almost all cases forest managers must be concerned with a mixture of marketed and non-marketed forms of production.

There are two main reasons why some forest benefits are not priced and sold, one technical and one political. The technical reason is that certain forest values are difficult to price and market in the usual way. The aesthetic value of a forest landscape, for example, would be difficult to parcel up and sell to individual consumers, and to exclude those who were unwilling to pay for it. In any event it would not be desirable to price it; a view of a forest landscape is an example of a true *public good* in the sense that its consumption by one consumer does not reduce its availability to others, so any positive price would inefficiently ration its consumption and reduce the value it generates. Other public goods are associated with the contribution of forests to the quality of the natural environment, such as air quality.

The political reason is that some forest products and services are not marketed because of public choice. For example, in contrast to the view of a forest landscape, access to recreational areas and campgrounds presents no technical obstacle to pricing. Indeed they are often priced by private owners. But governments frequently provide such facilities without charge. Sometimes a fee is charged for a general privilege to hunt or fish, but the charge is usually unrelated to any specific resources consumed, or even the amount consumed, and it is typically a nominal administrative fee rather than a market-determined price.

When benefits are not marketed, regardless of the reason, the problem of quantifying their value arises. The fact that they are not priced does not mean that they are valueless, of course, only that there are no market indicators of their value.

Thus the issue in this chapter is how to estimate the value of a forest product or service where users are prevented from expressing their evaluation of it by paying a price. In the absence of the usual market indicators of value, we must resort to indirect evidence about the demand for the product or service. As long as the full benefits accrue to individual consumers the problem is, specifically, one of finding and analysing other information in order to estimate how much consumers would be willing to pay for the good or service even though no price is actually charged.

Our interest in the economics of producing any good or service calls for attention to both the benefits and the costs—that is, the value of the product and the cost of producing it. However, the costs of producing unpriced goods and services are not usually very different in kind from those associated with the production of commercial products. In both cases the costs are usually reflected in expenditures for the labour, land, and capital needed to manage the forest for particular purposes. The difficulty lies on the benefit side, so the discussion in the following pages concentrates on the special problem of evaluating benefits wherever they are not priced.

CONSUMER SURPLUS AS A MEASURE OF VALUE

The value, or utility, that consumers gain from a good or service is reflected in their willingness to pay for it. Their willingness to pay indicates their willingness to give up other things, or income, in order to obtain that particular good, and so it measures their relative evaluation of it in money terms.

The willingness of consumers to pay for some article in a particular market is indicated by a market demand curve of the kind described in Chapter 3. Thus the problem of quantifying the value of a good or service focuses our attention on the demand curve for it.

Figure 9 illustrates a typical downward-sloping demand curve, dd′. If the product were sold in a market at a price op, the quantity purchased would be ox, and the total amount paid by purchasers would be the price multiplied by the quantity consumed, represented by the rectangle opp′x. But the value that consumers gain from consuming this good is equal to the price only at the margin. The triangle pdp′ indicates that some consumers would be willing to pay more than the price they all pay, op. And this amount—the amount consumers would be willing to pay in excess of the amount they actually pay—is referred to as *consumer surplus*. It is obviously part of the total value consumers gain from a good or service.

If a product or service is available at zero price, then *all* the value accruing to consumers is in the form of consumer surplus. In Figure 9, if the price were zero, consumption would increase to od′, the value gained by consumers would be the entire area under the demand curve odd′, and all this value would be in the form of con-

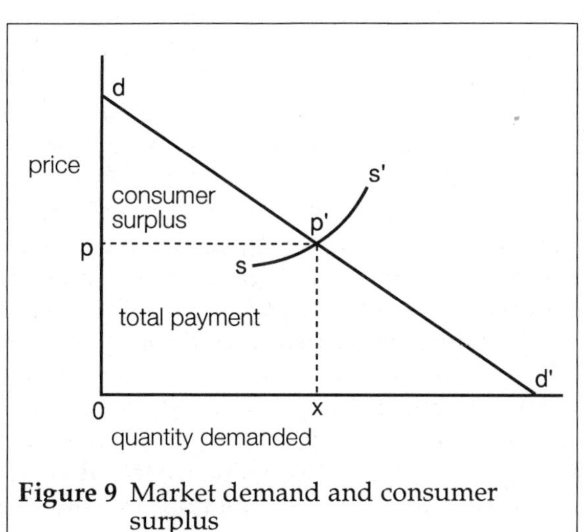

Figure 9 Market demand and consumer surplus

sumer surplus. So, to establish the value of a good or service provided without charge we must estimate the consumer surplus it generates, which in turn requires some way of defining its market demand curve and measuring the area under it.

As long as the product or service is sold to all consumers at the same price at least part of the total benefit accruing to them will be in the form of consumer surplus. If the price is zero, all the benefit will be in this form; if it is any positive amount, the total benefit will consist of both consumer surplus and payments to sellers. Sellers could eliminate consumer surplus altogether only if they could charge every buyer a different price, extracting from each the maximum amount he is willing to pay for each unit purchased. But sellers rarely have the opportunity to discriminate among buyers, and capture all the benefits in sales revenue, in this way.

The existence of consumer surplus obviously depends on a downward-sloping demand curve. If the demand were perfectly elastic, so the demand curve was horizontal at the price level p in Figure 9, there would be no consumer surplus. This is important, because it means that the need to estimate consumer surplus arises only when a significant change in the quantity of a good or service, or in its price, is being considered. If the quantity is marginal, consumer surplus can be ignored. For example, if we want to estimate the value of all the product consumed in a market, ox in Figure 9, the consumer surplus, pdp′, is a significant component of that value. On the other hand, if the task is to estimate the value of a marginal increment to the market supply of a good or service, such as that provided by one of many producers or that supplied from one of many tracts of land, the demand for that increment can reasonably be assumed to be perfectly elastic at the going price. Thus a small change at the margin of supply and demand will result in an insignificant change in consumer surplus, which can be ignored.

The latter is the usual case facing forest managers. They often want to know the benefit of providing for more recreation or some other value in a particular forest when this would add a relatively small increment to the total supply available to the relevant consumers. In such cases the potential change in consumer surplus may be negligible. But the larger the increment of supply relative to the total, and the more unique the character of the good or service, the less elastic the demand curve is likely to be and the greater the consumer surplus to be considered.

Sometimes an all-or-nothing decision must be made about the preservation of a unique feature of nature or a wilderness area. In these cases the total demand for the site must be evaluated.

EVALUATING UNPRICED RECREATION

To focus discussion of this issue, let us consider a particular type of unpriced benefit which is increasingly important in forest management: unpriced outdoor recreation. More specifically, we will consider the recreational value of a particular *site*. Forest managers and planners often have to decide whether to allocate a forest area to recreational purposes, so they need to estimate the value of the site if it were put to this use rather than to other uses.

Outdoor recreation is pursued in many forms, but most recreation experiences are a combination of anticipation, travel to and from the site, on-site activity, and recollection afterward. The demand for a particular recreational site is derived from the demand for the kind of recreational experience the site provides, just as the demand for timber is derived from the demand for final products made of wood, as explained in Chapter 3 and Figure 4.

As already noted, the value of a recreational site is reflected in the demand curve for it, which indicates the willingness of recreationists to pay for access to the site. Such a demand curve can be expected to take the form illustrated in Figure 9, but if access is free we can observe, at best, only the point d', the quantity demanded at zero price. The task therefore is to estimate the amount of recreation that would be consumed over the range of prices above zero.

Recreationists who are not required to pay a price for access to a forest for recreational purposes are likely nevertheless to incur costs when they take advantage of this opportunity. They usually incur travel costs to reach the site, and they may have to purchase supplies and equipment, among other things. Such expenditures do not measure the value they derive from a particular recreational opportunity or their willingness to pay for access to it. These expenses are analogous to the costs one might incur in attending a movie, such as the cost of gasoline, parking and baby-sitting services. What is needed is an estimate of the amount people would be willing to pay in the form of a toll to enter the particular recreational site, analogous to the theatre ticket.

The distinction between consumers' evaluation of something and the costs they incur in consuming it is important because the two are often confused. The expenditures of tourists and other recreationists are sometimes misrepresented as the value of the facilities that attract them. But recreationists' spending represents only the costs they incur, not the benefits they enjoy. The benefits of a recreational activity are reflected in the demand for it, not the demand for ancillary goods and services.

However, the expenditures of tourists and recreationists undoubtedly generate business and affect patterns of income and employment, so they provide useful information for assessing the economic impact of recreational opportunities in a region or local area. We can also use this information to estimate the net gain that accrues to others from this spending, by subtracting from the total amount spent the opportunity costs of the resources used in supplying the recreationists with the goods and services they purchase. But if we want to measure the value of particular recreational sites or facilities we must estimate the demand curves for them, which reflect the willingness of consumers to pay for access to specific recreational opportunities, quite apart from incidental expenses they incur.

Direct Techniques for Estimating Consumer Surplus

Direct techniques for estimating the value of unpriced recreational opportunities involve asking recreationists themselves for the information needed to specify relevant demand curves. Recreationists are simply asked to declare the maximum amount they would be prepared to pay to visit the site and participate in the recreation rather than go without it. This is usually done through mail surveys or on-the-spot questioning of recreationists or of a sample of them. Then, by arraying the responses from highest to lowest, the demand curve for the recreational opportunity can be drawn. The total consumer surplus is simply the sum of these responses, representing the area under the demand curve.

Many studies using this so-called "contingent valuation" method have been conducted to estimate the value of recreational sites such as parks and sport fisheries. If, say, a thousand recreationists per year indicate an average willingness to pay $50 for access to the site, the annual consumer surplus is estimated at $50,000. Using techniques described in Chapter 6, the present value of such an annual amount can be calculated to indicate the gross capital value of the recreational facility.

An alternative to asking recreationists how much they would be willing to pay for a recreational opportunity is to ask them the minimum payment they would accept to refrain from using it. Theoretically, this would yield a similar result as long as the values involved are insignificant in terms of the consumers' total income. However, the "income effect," which explains that a dollar reduction in income is weighted more heavily than a dollar increase in income, ensures that the minimum acceptable bribe to abstain will never be less than the maximum willingness to pay, and some surveys have found it to be much greater.

These two approaches yield different measures of consumer surplus. The maximum willingness to pay in addition to what a consumer already pays for a good is called the *equivalent variation* measure, while the minimum acceptable compensation to do without it is the *compensating variation*. The former is often considered to be more appropriate for evaluating a proposed project while the latter is more suitable for assessing the value of an existing recreational resource.

The difficulty with these direct techniques is in obtaining rational and consistent expressions of value from recreationists in response to hypothetical questions about their willingness to pay to use a recreational resource that they actually use without charge. The emotion often attached to freely accessible recreation, coupled with suspicions about the purpose of such questions, is likely to produce answers that are deliberately or subconsciously distorted (such as "it is priceless" or, at the other extreme, "I would refuse to pay anything"). Moreover, if the value declared by any recreationist were actually charged, he would probably alter somewhat the quantity of the recreation he consumes, which also complicates the interpretation of results.

Indirect Measures of Consumer Surplus

The practical limitations of direct techniques have led to the development of econometric methods for inferring recreationists' willingness to pay from indirect evidence drawn from their observed behaviour. These approaches can be best explained by sketching, first, their theoretical underpinning.

Theory of recreation behaviour. Some of the costs of participating in a recreational experience are *fixed* costs in the sense that they do not vary with the quantity of recreation consumed as measured in recreation days spent at the site. These fixed costs include the cost of travelling to and from the site and any fee that must be paid for access. Other costs, such as the cost of food and supplies, are *variable* costs because they depend on the duration of the recreational experience.

The fixed and variable costs incurred by an individual recreationist are depicted in Figure 10, in which the recreationist's income is measured along the vertical axis and the number of on-site recreation days consumed is indicated on the horizontal axis. Suppose his total income is equal to the vertical distance OY, and to participate in the relevant recreational opportunity at all he must incur fixed costs equal to T_xY, reducing his remaining income to OT_x. His variable or on-site costs of recreation are represented by the slope of the line T_xR_x (which is shown as a straight line on the assumption that the marginal cost of a recreation day is constant). The kinked line YT_xR_x thus traces the recreationist's opportunities for dividing his income between this recreational activity and other things.

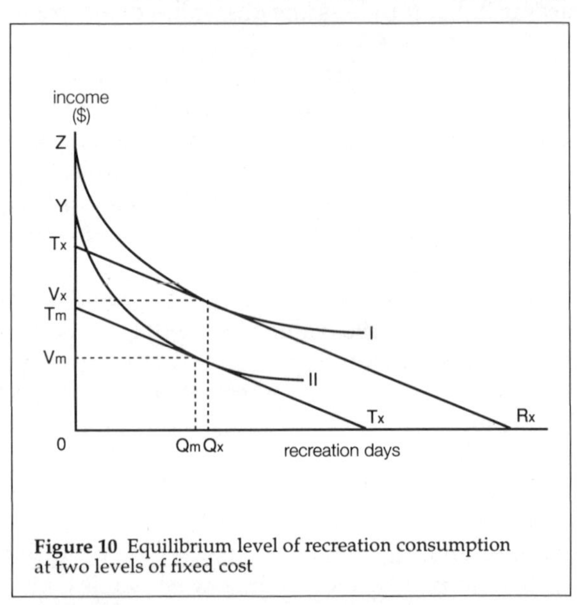

Figure 10 Equilibrium level of recreation consumption at two levels of fixed cost

The various combinations of recreation and income that will yield the recreationist the same utility or satisfaction can be represented by indifference curves. One such curve, curve I in Figure 10, is just tangent to the line tracing his market opportunities, and thus represents the highest level of satisfaction he can attain, by consuming OQ_x recreation days. By doing so, he will spend V_xY on the recreation, of which T_xY are fixed costs and V_xT_x are variable costs.

This recreationist clearly would be willing to pay more for access to the recreational opportunity than the fixed cost T_xY that he incurs. At most, he would pay fixed costs of T_mY, which would leave him at the same level of indifference as he would be by consuming no recreation at all. This is shown by indifference curve II, which runs through point Y, indicating his level of satisfaction without any recreation. Note that if he had to incur this extra fixed cost he would spend a total of V_mY on recreation and reduce his consumption of it to OQ_m.

This depiction illustrates the two measures of consumer surplus identified earlier. The amount T_mT_x is the equivalent variation, the maximum that the recreationist would pay to gain access to the site in addition to the costs he actually incurs. The amount YZ is the compensating variation, the minimum amount he would need to have added to his income in order to leave him equally satisfied without participating in the recreation. In the following paragraphs we adopt the former, more traditional definition of consumer surplus and show ways of estimating it.

Estimating willingness to pay. The value of a recreational resource enjoyed by a recreationist under free access—that is, his consumer surplus—can be expressed as the maximum access fee or toll he would be willing to pay in addition to his fixed costs. As long as he would react to a toll in the same way as he reacts to fixed costs, the maximum toll he would pay is T_mT_x in Figure 10. The sum of these maximum hypothetical tolls for all participating recreationists is the total consumer surplus generated by the recreational opportunity, or the area under the demand curve for it.

The simplest way to estimate this willingness to pay is based on survey information about the travel costs incurred by the participating recreationists in travelling to and from the recreation site, which depend mainly on the distance they travel. By assuming that all participating recreationists have the same willingness to pay, and that the one who incurs the highest travel cost is marginal in the sense that he would not be willing to pay anything more, the consumer surplus of all the rest can be expressed as the difference between the travel cost each incurs and the travel cost of the marginal participant. Ranking these estimates of individual willingness to pay from the highest to the lowest can be expected to reveal a downward-sloping demand curve like that in Figure 9.

Thus, if a survey of the users of a camping area revealed that the camper who incurred the highest travel cost spent $40 travelling to and from the site and the average of all campers was $15, the average consumer surplus is estimated at $25 by this method. Multiplying this figure by the total number of campers who visit the site per year provides an estimate of the annual value of the recreational facilities.

This procedure involves some tenuous assumptions: that all the recreationists are equally willing to pay for the recreation regardless of differences in their incomes and other characteristics; that their cash outlays for travel fully represent the costs they incur to obtain access to the site; that they would respond to a toll in the same way as they respond to travel costs; and that the participant who incurs the highest travel cost is marginal. These assumptions are not likely to conform with reality, and they obviously limit confidence in such estimates.

The assumption that all recreationists are equally willing to pay is particularly questionable because we normally observe larger numbers participating at relatively low costs, and fewer at high costs, consistent with the normal downward-sloping demand curve. To assume that all would be prepared to pay the same amount as the one who paid the most almost certainly exaggerates the consumer surplus.

An alternative technique that avoids this assumption involves measuring the sensitivity of recreationists to the costs they must incur to reach the site, and using this information to estimate their response to an access fee. The method is illustrated with a simple example in Figure 11. The geographic area from which recreationists are drawn to a recreation site is divided into concentric zones, within each of which travel costs to the site are approximately uniform. For each zone, the total population, the number who visit the recreation site, and their average travel cost to and from the site are assessed, as shown in the first four columns of the table in Figure 11. This figure provides the information for establishing the relationship between the participation rate and travel costs for the entire population, as shown in quadrant A.

The second quadrant uses this relationship to estimate the reduction in the number of visitors to the site that would result from adding an access fee, of varying amount, to the travel cost of all visitors. The table shows the calculation for a hypothetical access fee, or price, of \$5 and \$10. The relationship in quadrant A gives the lower participation rate that could be expected from each zone when the price is added to the travel cost, and that rate is applied to the total population of the zone to yield the number of participants who could be expected at that price. These estimates are then used to plot the demand curve for the recreation site, as in quadrant B.

In this simplified example the demand curve is a straight line, and so the total consumer surplus represented by the triangle under the demand curve can be easily calculated at \$442,500 (i.e., $.5 \times \$30 \times 29{,}500$).

Because it depends on estimates of participation rates from whole populations, this technique becomes less reliable when applied to recreational sites that attract visitors from a wide area, encompassing large populations having different recreational alternatives.

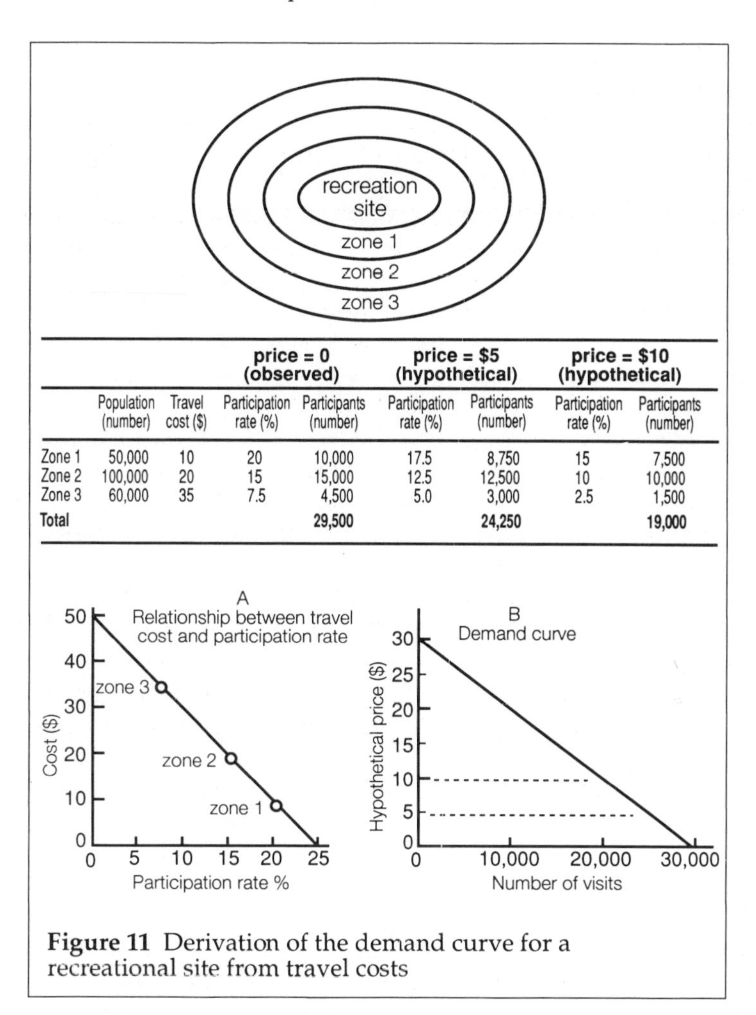

			price = 0 (observed)		price = $5 (hypothetical)		price = $10 (hypothetical)	
	Population (number)	Travel cost ($)	Participation rate (%)	Participants (number)	Participation rate (%)	Participants (number)	Participation rate (%)	Participants (number)
Zone 1	50,000	10	20	10,000	17.5	8,750	15	7,500
Zone 2	100,000	20	15	15,000	12.5	12,500	10	10,000
Zone 3	60,000	35	7.5	4,500	5.0	3,000	2.5	1,500
Total				**29,500**		**24,250**		**19,000**

Figure 11 Derivation of the demand curve for a recreational site from travel costs

Other techniques have been devised to recognize differences among recreationists that influence their willingness to pay. For example, using data obtained through surveys, those who utilize the relevant recreational resource can be classified by income, which can be expected to be a primary determinant of their willingness to pay for it. Within each income group they can then be ranked according to the fixed costs they incur. If there is a sufficient sample in each income group, it can be assumed that the recreationist who incurs the highest fixed cost is at or near the margin, in the sense that he enjoys no consumer surplus (corresponding to one who incurs fixed costs of T_mY in Figure 10). Then it is necessary to assume only that the recreationists who have similar incomes and participate in the same recreational activity are equally willing to pay for it.

Given these assumptions, the consumer surplus enjoyed by each recreationist is equal to the difference between the fixed cost he incurs and the fixed cost of the marginal (highest cost) recreationist in the same income group. The sum of these differences for all recreationists in all income groups is the total consumer surplus generated by the recreational opportunity.

To obtain the demand curve for the recreational facility from this information, the number of recreationists who would participate at any particular price can be estimated as those for whom the difference between their fixed cost and that of the marginal recreationist in their income group exceeds the hypothetical price.

This technique is capable of recognizing differences in income and possibly other characteristics of recreationists such as education, age, and family circumstances that are likely to influence their willingness to pay. However, some of the assumptions on which the estimates are based may give rise to error, especially the assumptions that all recreationists within a defined group are equally willing to pay for the activity, that the one who incurs the highest cost is marginal, and that all recreationists face similar alternatives.

The Hedonic Method

A quite different approach to evaluating recreational resources is based on recreationists' responses to the quality characteristics of different sites. In contrast to the techniques discussed above, which attempt to evaluate a recreational resource in isolation, the *hedonic method* uses information about the characteristics of a variety of sites to estimate the value recreationists ascribe to them.

Each site is viewed as a bundle of characteristics of importance to recreationists; for example, the characteristics for a sportfishing site may be fish abundance, fish size, privacy, clean water, scenic beauty, and so on. Every sportfishing opportunity accessible to a consumer has a different combination of these qualities. The hedonic framework uses observations about how recreationists choose among alternative sites to estimate implicit prices, or the value they attach to particular characteristics.

The technique involves analysing the relevant characteristics of a variety of recreational sites and attaching a value to each characteristic using a numerical scale. The costs incurred by recreationists who visit the sites are regressed against the site attributes to yield a "shadow value" for each. The result is an estimate of recreationists' willingness to pay for each attribute. Any particular site can then be evaluated in terms of its quality characteristics.

An attractive feature of the hedonic method is its specific recognition of the quality of recreational opportunities, and its ability to explain differences in the value of different sites on the basis of their quality characteristics. This can provide valuable information for resource planners. However, the technique presents some practical difficulties in scaling the quality of site attributes and in calculating the values attached to them.

The hedonic method, like other methods of estimating consumer surplus that draw inferences from recreationists' responses to costs, implies an assumption that the recreationists' sole purpose in incurring their costs is to participate in the recreational opportunity in question. This is probably a realistic assumption for certain types of recreation such as hunting or fishing. But campers and tourists are likely to be less single-minded about their recreational objectives, and if they casually visit a particular area in the course of a wide-ranging tour it would obviously be inappropriate to attribute all their travel costs to that single experience. When costs are incurred for multiple purposes some method of prorating them is called for.

A related implication of these techniques is that the consumer's travel costs are fully measured by his cash outlays, and specifically that he does not consider as a cost the time he must expend in travelling to and from the site. In reality, these circumstances vary. A family on a camping trip may consider the travelling part of the recreational experience and therefore the time involved appropriately ignored in accounting for costs. In contrast, hunters or fishermen are likely to regard time spent travelling as an encroachment on their on-site recreation, so some cost should be ascribed to it. Some of these problems can be illuminated by careful attitude surveys of recreationists.

INTERMEDIATE PRODUCTS IN FOREST RECREATION

Outdoor recreation is consumed directly by consumers, so its value is properly assessed in terms of consumers' willingness to pay for it, as described above. However, forest managers are often concerned with providing less direct human benefits in the form of wildlife, fish, amenity, and general quality of the natural environment. Such benefits add another dimension to the problem of evaluation.

For purposes of economic evaluation, it is essential to identify clearly the nature of the human values created when forests are manipulated for a particular purpose. Fish and wildlife, for example, are usually valued for their contribution to the recreational quality of a forest, either as objects of viewing or of hunting and fishing. There may be exceptions, where wildlife is managed for commercial, scientific, or purely environmental purposes, but for present purposes we shall assume that the value generated is in the form of recreational fishing, hunting, or viewing, as is more often the case.

Wildlife, valued by hunters, fishermen, and sightseers, is thus an intermediate product in the production of recreation, in the same sense that timber was described as an intermediate product in the production of housing and newspapers in Chapter 3. This observation is important because the issue is often confused in attempts to ascribe values directly to fish and game or to the harvest of hunters and fishermen. These efforts miss the point that the ultimate product of fish and game management is not the fish and game themselves but the recreation they support. More of them will increase the quantity or quality, and hence value, of the recreational opportunity. But a bagged bird or deer or fish must be regarded primarily as a by-product of a recreational experience. A fish may or may not have intrinsic value to the fishermen (who may release it, eat it, or give it away). Some chance of catching a fish is essential for a fishing experience, but the fact that fishermen can enjoy fishing without catching anything indicates that the harvest is only one of many dimensions of the quality of a fishing experience.

Additional fish or game can increase the recreational value of a forest in either or both of two ways. One is through enhancing the quality of the recreational experience. More wildlife will usually result in more sightings by nature lovers, and more fish or game will increase the success rate of sport fishermen and hunters. Such improvements in the quality of the experience will increase the willingness of recreationists to pay for it, shifting upward the demand curve for the recreational opportunity.

The other effect, resulting from the increased attractiveness of the recreational opportunity, is that more recreationists will participate in it if access is free. This implies a shift of the base of the demand curve to the right, again increasing the area under the demand curve. A sufficient increase in the number of users could forestall any improvement in the average harvest that would otherwise result from the enhanced resources. In any event, through either or both of these effects, the demand curve will shift, and the value of the additional fish or wildlife is reflected in the increase in the area under it, which is the increment in consumer surplus enjoyed by hunters or fishermen.

RECREATIONAL CAPACITY, QUALITY, AND CROWDING

These considerations illustrate the interdependence between the quality of a recreational experience and the level of consumption of it. Both of these affect its value.

One dimension of the quality of a recreational opportunity is congestion, or the degree of crowding. If the demand for an unpriced recreational opportunity increases sufficiently, and pricing or other means of rationing access are unavailable, the site will become crowded, reducing the quality and value of the recreational experience.

The effect of crowding is illustrated in Figure 12. At zero price, increased numbers of recreationists shift the base of the demand curve to the right. However, the lower quality of the recreational experience means that the recreationists will not be willing to pay as much for it, hence the demand curve is lower. Whether such a change will increase or decrease the total value generated, represented by the area under the curve, will depend on its shape and the degree of these opposing effects. In Figure 12 the lower triangle between the two curves measures the gain in consumer surplus from additional numbers of consumers; the upper triangle indicates the loss resulting from the lower quality of the more crowded site.

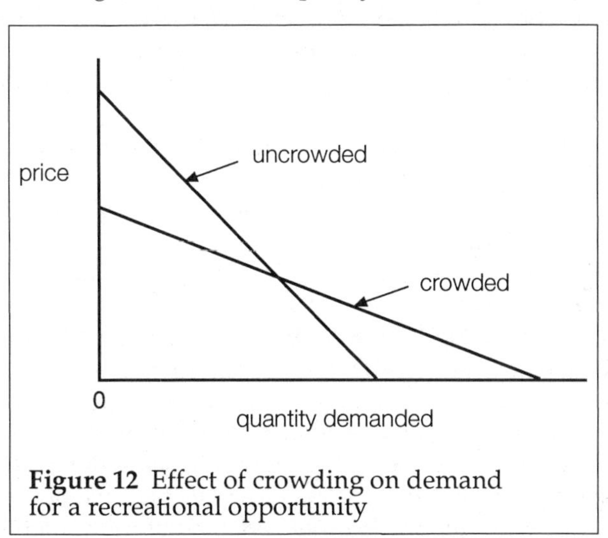

Figure 12 Effect of crowding on demand for a recreational opportunity

It should be noted that crowding is not always considered detrimental; a "good crowd" can enhance the appeal of recreational facilities such as ski resorts or holiday camps. But for most forms of forest recreation solitude enhances quality, and crowding conversely diminishes the value of the experience.

Both the quality and the capacity of a recreational site can usually be enhanced through investment. The *quality* of a recreational resource can be improved by such measures as grooming campsites and trails, enlarging wildlife populations, or refraining from unsightly timber harvesting practices. The effect of such quality improvement, in terms of the conceptual framework above, is to shift the demand curve for the recreation to the right, indicating larger numbers of demanders at zero price, and upward, indicating their greater willingness to pay for the opportunity.

The *capacity* of a recreational site can be expanded by investment in additional campsites, trails, and so on, so that more recreationists can be accommodated without diminution of the quality of the experience. Expansion of capacity therefore can accommodate a shift to the right of the base of the demand curve without reducing participants' willingness to pay, represented by the height of the curve.

In all these cases the benefit resulting from the investment is the increase in the area under the demand curve, which can be estimated by one of the techniques described earlier in this chapter. It is this benefit which must be compared with the cost of the improvement to measure its net gain.

If a recreational site were operated by a profit-maximizing owner who was free to price access, he would, given the capacity and quality of the site in the short run, select that combination of price and number of recreationists that would generate the greatest net revenue. In the long run he would invest in the site's capacity and quality whenever the cost was less than the resulting increase in revenue.

This behaviour would ensure that crowding would not diminish the value of the resources. But where access is free and uncontrolled this sort of optimization is not possible, and crowding may erode the value of the recreation. Thus it is commonplace, where a policy of free access to recreational resources applies, to find examples of campsites, wilderness areas, and fishing and hunting opportunities becoming so crowded that their attractiveness to potential users is diminished.

In the absence of pricing, the erosion of recreational values through crowding can be prevented through other means of rationing access, such as by allowing a limited number of users on a first-come-first-served basis, by drawing lots or some other scheme for admitting only some of the demanders. These other techniques are less effective in generating the maximum value from the recreational opportunity because, unlike pricing, they cannot ensure that those who value it most highly will have access to it. They can, however, protect the quality of recreational sites from deteriorating through crowding.

EXTERNALITIES

So far, we have been referring to benefits that accrue to consumers, so that the value of the good or service can be assessed in terms of the potential willingness of consumers to pay for it. However, in some cases benefits or costs accrue to people other than the consumers, giving rise to *externalities* of the kind described in Chapter 2. These external values are sometimes significant, and it is necessary to estimate them to supplement estimates of the values that accrue to consumers.

Sometimes external costs and benefits are manifested in financial terms. For example, a forest or park not only provides benefits to its users but may also increase the value of adjacent private properties. In this case the external benefit accruing to local landowners is simply the enhanced property values.

More often, external benefits are not reflected in any market prices, and estimating their value is more problematical. Several kinds of such external effects are important in forest management and it is helpful to distinguish among them.

Option value. People who do not participate in a recreational experience or seek out a particular amenity may nevertheless value the opportunity to do so. They are thus willing to pay something for preserving the option for themselves or their children, even though they are not among the active consumers. The value people put on preserving such opportunities is *option value*.

Option value is particularly relevant to decisions about unique features of nature where irreversible decisions are being considered. For example, evaluation of a proposal to build a hydroelectric reservoir that would eliminate a unique wilderness should take account of the option value associated with the wilderness, in addition to any current recreational or aesthetic value it generates. Where the relevant resources are not unique, or can be replaced, which is more often the case in managed forests, this special value is less important.

Preservation value. A closely related concept is *preservation* or *exis-*

tence value, which is the value people put on something regardless of any interest in direct consumption. For example, many people are willing to pay something to preserve the whooping crane, the woods buffalo, and a rainforest even though they have no expectation of seeing these things. Their support for public expenditure on protecting them is evidence of their willingness to pay simply to maintain their existence.

Public goods. Early in this chapter it was noted that forests sometimes yield true *public goods*, which suppliers cannot parcel up and sell to those who are willing to pay a price and exclude others, and which consumers can consume without diminishing the supply available to others. The classic example of a true public good is a lighthouse; forest examples include the amenity of a landscape and general environmental quality.

These various types of externalities are not always easy to separate. The value of maintaining a forest may be a mixture of consumer surplus, option value, and preservation value, and public goods may generate values in all these forms. They are all exceedingly difficult to evaluate except in the rare cases where market indicators of value are available. Estimates of their value must rely on subjective assessments or on public surveys designed to establish the willingness of people to pay for them rather than go without them.

COST-EFFECTIVENESS

As noted at the beginning of this chapter, the costs of providing some unpriced goods and services are often much easier to estimate than the benefits. For some purposes forest managers find it sufficient, instead of attempting to estimate the value of certain unpriced forest benefits, to compare the cost of providing them in alternative ways or places.

This is an expedient technique where the resource planner's objective is to produce a certain quantity of fish, or game, or camping opportunity in a region. He can compare the cost of all the alternative ways of contributing to the goal, and by choosing the least-cost means he will meet his objective at lowest possible aggregate cost.

Obviously, such analyses of cost-effectiveness reveal nothing about the value of the benefits produced, which is the purpose of all the techniques discussed earlier in this chapter. It is, nevertheless, a helpful technique for ensuring consistency and efficiency in meeting predetermined forest management objectives.

PRACTICAL LIMITATIONS

In making these assessments of unpriced values it is particularly important to clarify the type of benefit or benefits being generated, those who are the beneficiaries, and how the chosen valuation technique captures these values. This is because the non-marketed products and services of forests are often difficult to define and measure, as noted above, and they often generate emotions which can confuse or distort the information needed to assess them.

It is worth noting that the evaluation problem discussed in this chapter is not due to the "intangibility" of natural values, as is often suggested. The enjoyment people derive from a painting, or a record album, or a book of poetry is intangible in the usual sense, but the value of these things is nevertheless observable in market prices. It is because certain forest values are unpriced, not because they are intangible, that their valuation is difficult.

The difficulties of estimation suggest that analysts should take advantage of market indicators of values wherever they are available. For example, external costs and benefits of the kind encountered in forestry often become capitalized in property values. And sometimes the value of unpriced recreational facilities such as campgrounds can be estimated with reference to the prices charged by comparable facilities that are priced.

In this chapter we have considered a variety of non-marketed values that may be produced from forests, and various approaches to evaluating them. Each of the direct and indirect techniques for estimating consumer surplus and other unpriced values has its special strengths and weaknesses which make it most suitable for use in particular circumstances.

None of these methods provide more than rough estimates of values. But they can provide useful guidance in deciding how forests can be best used and developed, which otherwise must be resolved through guesswork. Techniques for evaluating unpriced goods and services continue to be developed, and accumulating experience in applying them is leading to improvements in the reliability of estimates. But they are rarely precise, and the results must be used cautiously.

REVIEW QUESTIONS

1 The right to fish in a particular stream or to hunt in a particular area is usually priced in Europe but not in North America. What are the reasons for this difference? How does it affect the problem of evaluating these recreational resources?
2 What is "consumer surplus"? With reference to a typical demand curve, explain the relationship between consumer surplus and the total amount consumers are willing to pay for a product when the price is (a) positive, and (b) zero.
3 Why is it inappropriate to use the expenditures of visitors to a park as a measure of the value of the park?
4 Recalculate the total consumer surplus that would be attributable to the recreation site depicted in Figure 11 if the participation rate for each of the three zones were doubled.
5 How does the number of game animals in a forest affect the demand for hunting? What other factors are likely to affect the demand and value of hunting in the forest?
6 Why are "option value" and "existence value" likely to be more important in assessing the economic implications of harvesting a virgin stand of old-growth Douglas-fir than the harvesting of a Douglas-fir plantation?

FURTHER READING

Bishop, R.C. 1982. Option Value: an exposition and extension. *Land Economics* 58(1):1–15

Bowes, Michael D. and John V. Krutilla. 1989. *Multiple-Use Management: The Economics of Public Forestlands*. Washington, D.C.: Resources of the Future. Chapter 7

Boyd, Roy G., and William F. Hyde. 1989. *Forestry Sector Intervention: The Impacts of Public Regulation on Social Welfare*. Ames: Iowa State University Press. Chapter 8

Brown, Gardner, Jr., and Robert Mendelsohn. 1984. The hedonic travel cost method. *The Review of Economics and Statistics* 66(3):427–33

Cesario, F.J., and J.L. Knetsch. 1976. A recreation site demand and benefit estimation model. *Regional Studies* 10(1):97–104

Clawson, Marion, and Jack L. Knetsch. 1966. *Economics of Outdoor Recreation*. Baltimore: Johns Hopkins University Press for Resources for the Future. Part 2

Davis, L.S., and K.N. Johnson. 1987. *Forest Management.* 3rd ed. New York: McGraw-Hill. Chapter 12

Freeman, A. Myrick III. 1979. *The Benefits of Environmental Improvement: Theory and Practice.* Baltimore: Johns Hopkins University Press for Resources for the Future. Chapters 5, 6, and 8

Krutilla, John V. 1970. Evaluation of an aspect of environmental quality. In *Proceedings of the Social Statistics Section of the American Statistical Association.* 13th annual ed. Papers presented at the annual meeting of the American Statistical Association, Detroit, Michigan, 27-30 December 1970 under the sponsorship of the Social Statistics Section. Washington, D.C.: American Statistical Association. Pp. 198-206. See also "Discussion," pp. 215-16

—, and Anthony C. Fisher. 1985. *The Economics of Natural Environments: Studies in the Valuation of Commodity and Amenity Resources.* Rev. ed. Washington, D.C.: Resources for the Future. Chapters (to come)

Pearse, Peter H. 1968. A new approach to the evaluation of non-priced recreational resources. *Land Economics* 44(1):87-99

Smith, V. Kerry, and Yoshiaki Kaoru. 1987. The hedonic travel cost model: a view from the trenches. *Land Economics* 63(2):179-92

Walsh, Richard G. 1986. *Recreation Economic Decisions: Comparing Benefits and Costs.* State College, PA: Venture Publishing. Chapters 5-8

—, John B. Loomis, and Richard A. Gillman. 1984. Valuing option, existence, and bequest demands for wilderness. *Land Economics* 60(1):14-29

CHAPTER FIVE

Land Allocation and Multiple Use

Out in the Forest . . .

To get the most out of Peavey Forest Products Limited's forestland, Ian Olson has a silvicultural plan for each stand. But even on the best land there is a limit to the amount of silvicultural effort that is profitable. Finding that limit is one of the forester's most complicated tasks because he must weigh the cost of more silviculture against his estimate of the value of the extra timber that would result. The answer differs for each stand, and it keeps changing as silvicultural costs and timber values change.

The amount of land that can profitably be used for timber production also depends on timber prices and costs, and it is affected as well by the value of the land in other uses. In recent years, as returns to farming on marginal land have declined, the company has turned some abandoned farmland into softwood plantations. On the other hand, it lost some forestland to growing recreational demands, when the government purchased a waterfront area and added it to the public park.

Nowadays, nearly all the company's forestland is capable of generating non-commercial values in addition to timber. Some of these, such as wilderness values, are incompatible with timber production on the same site. But most, including the important recreational and aesthetic values, conflict only to a limited degree, and combined production of two or more benefits is possible.

Olson finds it increasingly necessary to allow for these competing and complementary values in designing multiple resource use plans. He must identify the various uses and combinations of uses that are possible on each site, and the relative values they generate, in order to plan the most beneficial pattern of land use.

The most fundamental decisions in forest management relate to the allocation of land among alternative uses. The economic challenge of this issue is to identify the use or combination of uses that will generate the greatest value. This chapter reviews the economic and technical relationships that determine the optimum pattern of land use.

Land usually can be used for a variety of purposes. Choosing the most productive form of land use calls for attention to economic efficiency at two levels. One is the most efficient way to manage land for any particular purpose; that is, how to apply labour and other inputs to generate the maximum return under a given land use. The second involves selecting among the alternative uses to find the one, or the combination of uses, that yields the highest net return. We turn first to the problem of maximizing the return under a particular form of land use.

INTENSITY OF LAND USE

Efficient production involves combining various factors such as land and labour in such a way as to generate the maximum possible net return. Where one factor in the production process is fixed in supply, maximizing the aggregate net return involves maximizing the return to the fixed factor, after the costs of all the variable factors have been met. In forest production land is usually the fixed factor.

The problem of efficient use of a tract of forest land is therefore one of identifying how much of the other productive factors such as labour and capital can advantageously be applied to it in the forest production process—that is, how much labour and capital must be used to generate the maximum *land rent*.

This is, in other words, the question of the optimum *intensity of forest management*. Assuming that the forest manager will use the factors of production available in the most productive way that technology allows, we must identify the quantities of each that must be combined to generate the maximum return to the forest land.

The solution to this problem lies in the conditions for economic efficiency reviewed in the Appendix to Chapter 2. There it is noted that each factor in production should be employed up to the point at which the return it generates at the margin declines to its cost.

This condition is illustrated in Figure 13 in terms of the efficient amount of labour to employ on forest land of specific quality. The curve in the upper quadrant shows how the value of a forest crop can be increased if more labour is applied to it, increasing the intensity of silviculture. The curve in the lower quadrant shows the corresponding *marginal revenue product* of labour; that is, the additional value generated by an increment of labour over the range of labour

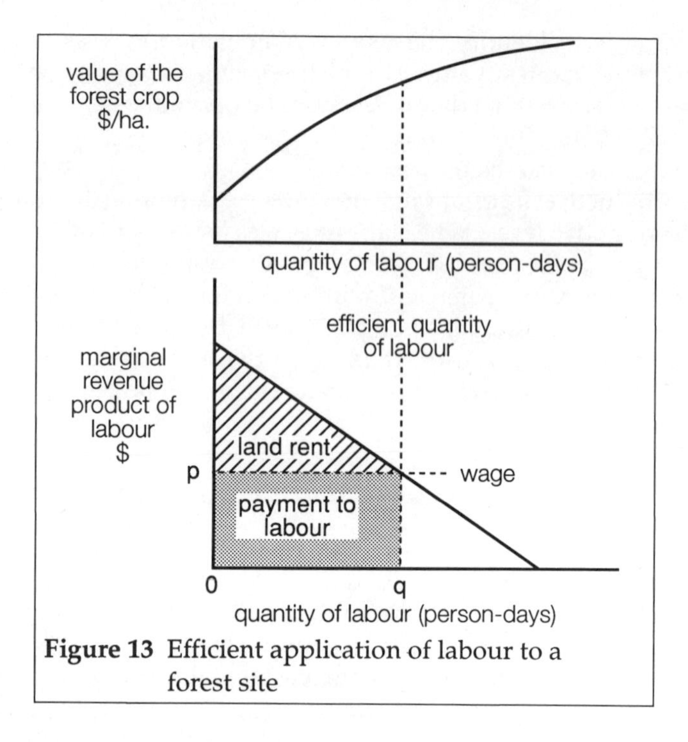

Figure 13 Efficient application of labour to a forest site

intensity. As explained in Chapter 2, this incremental value declines as more labour is added because of diminishing returns, reflected in the diminishing slope of the curve in the upper quadrant.

Given a price of labour, or wage rate, op, additional labour can be advantageously employed, contributing to the land rent, up to the quantity oq, at which point its marginal revenue product and cost are equal. This point defines the *intensive margin* of land use. At this level of employment labour can make its maximum contribution to the land rent, illustrated by the upper triangle between the curve and the price line in Figure 13. This is therefore the efficient combination of labour with land of this quality, and a corresponding relationship applies to all other variable factors of production.

Each forest site has unique qualities of fertility, location, and terrain that determine its productivity and its response to other factors of production. Usually land of lower productivity will yield a lower return to labour, implying a lower curve in the upper quadrant in Figure 13. But the efficient quantity of labour to apply to the land is determined by the *marginal* revenue product of labour, or the *slope* of the curve in the upper quadrant. The marginal revenue product of labour and other individual factors of production is less closely correlated with the inherent productivity of land.

A lower or more rapidly diminishing marginal revenue product of labour would indicate that a smaller amount of labour could be advantageously employed, and its contribution to the land rent would be smaller as well. Similar relationships apply to other inputs in forest cultivation, so the lower the responsiveness of the forest to silviculture, the lower is the potential land rent.

These relationships determine the optimum intensity of forest management. Generally, the higher the productivity of the land the higher the potential land rent, and the greater the quantities of other factors of production needed to generate it. This theory explains why land of high productivity usually warrants more intensive silviculture and other forms of forest management than land of low productivity.

EXTENSIVE MARGIN OF LAND USE

In Chapter 3 we examined how the supply and demand for timber determine its equilibrium market price, and noted that an increase in demand would result in a higher price and greater production. The increased production would result from the incentives of producers to cultivate forest land more intensively, as explained above, and to bring more land into forest production.

The latter effect is illustrated in Figure 14, which builds on Figure 6 of Chapter 3. The upper quadrant shows the equilibrium price and annual production of timber resulting from the interaction of market supply and demand. The lower quadrant shows, for each possible price and corresponding production level, how much land can profitably be used for timber production.

If the land has no other productive use, the most productive land will be employed in timber production at low timber prices, and at higher prices progressively less productive land will be drawn into production. At the price op in Figure 14 the equilibrium level of production is oq, and ol land can generate a return under timber production. Of this land, the most productive will earn a rent, as illustrated in Figure 13. The poorest, or marginal land, will earn no rent, which can be illustrated in Figure 13 by a lower curve of marginal revenue product that intersects the vertical axis at point p. Figure 14 shows how an increase in demand will increase both the equilibrium price and level of production and attract more land into timber production.

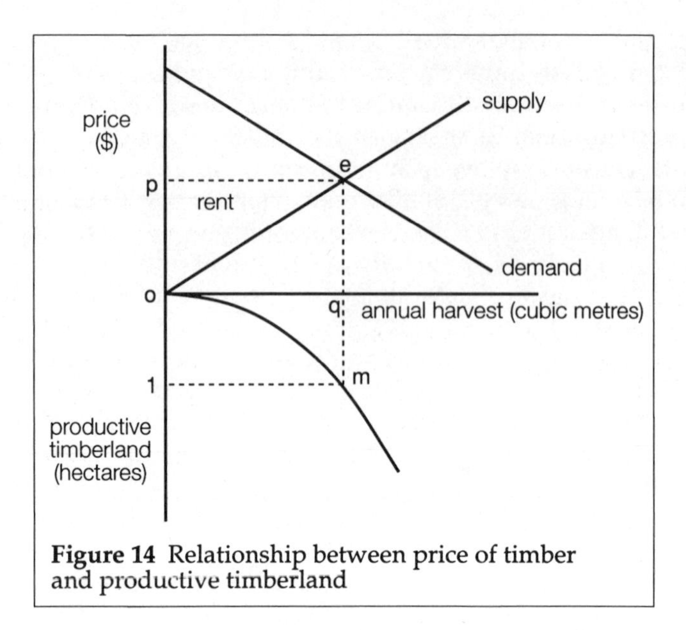

Figure 14 Relationship between price of timber and productive timberland

ALLOCATION AMONG USES

This representation of the efficient allocation of land to timber production becomes more complicated when the land can be used in a variety of productive ways. Choices often have to be made about whether to allocate rural land to agriculture, forestry, recreation, or some other use when each can generate a net return. The allocation of land among its alternative uses therefore is the second dimension of efficient land use.

We have already discussed how other factors of production are applied to land in order to generate maximum land rent. Assuming that the way to do this for all possible uses of the land is known, the task of efficient allocation of land among uses is to select the use that will generate the greatest rent in any particular time and place.

The capability of land to generate economic returns depends on a variety of factors: its fertility, its distance from markets, its topography and accessibility, and so on. The importance of each factor differs for different uses. The quality of land, or the determinants of its economic potential, can thus be viewed as a bundle of characteristics which have varying importance for different uses.

To illustrate the problem of allocating land among competing uses let us consider one such characteristic in isolation: distance from an urban centre. Since proximity to an urban centre is an important dimension of the quality of the land, we find land close to urban centres used more intensively than land at great distance from them, as the preceding theory suggests. But proximity to an urban centre is a more important dimension of the quality of land for some purposes than for others: it contributes more to the potential productivity of land for commercial purposes than for forestry, for example. So land at greater distance from urban centres is not only used less intensively in any particular use but is also used for different purposes.

This pattern is illustrated in Figure 15. Assuming all other characteristics of the land are identical, the figure shows how the potential land rent under various uses declines at progressively greater distance from an urban centre. In all uses, the value of the land is greater the closer it is to the urban centre, but the use that can generate the highest rent varies over the spectrum. Commercial use, which is the most intensive, yields a higher return than all others at the urban centre and forestry, the least intensive activity, makes the most productive use of the most remote lands. Farming generates the highest rent over an intermediate range, cd.

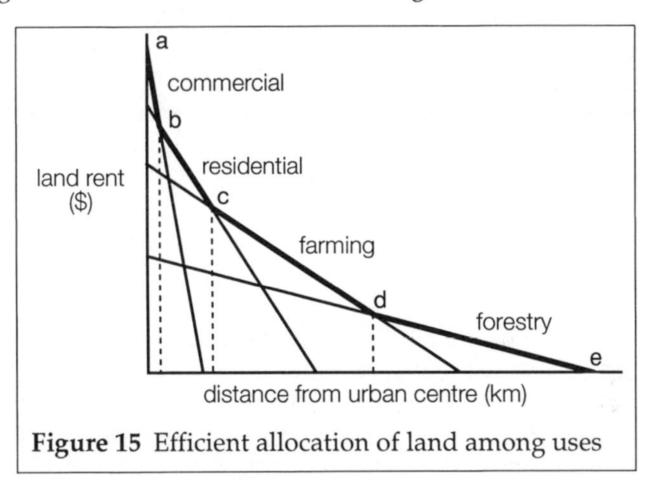

Figure 15 Efficient allocation of land among uses

Thus we observe concentric patterns of land use around urban centres. But distance from an urban centre is only one of many quality characteristics of land that determine its capacity to generate rent. Its fertility, topography, and many other dimensions similarly influence its relative value in different uses, and all these qualities blended together result in complicated patterns of efficient land allocation.

This representation helps to explain a number of other important observations about land use as well. First, it shows that land can often generate returns in more than one use. The use that generates the highest returns is the most efficient, while the rent in its next highest use represents the *opportunity cost* of the land, or the value it could generate in its next-best use. The *differential rent* is the rent the land earns in excess of its opportunity cost. Unlike the case in which land has no alternative use, land capable of earning rent in other uses can be efficiently allocated to timber production only if the rent it earns in timber production exceeds the rent under the alternative uses.

Second, it is not always efficient to allocate to a particular use the most productive land for that use. Figure 15 illustrates that land closer to the urban centre can earn a higher return in farming than land within the range cd, but it generates even higher returns in residential and commercial uses.

Third, the allocation of land to its highest use ultimately depends on the value of the outputs and the cost of the inputs in each alternative use, and because these constantly change so does the efficient allocation of land. Changes in technology, costs, and prices continuously shift the boundaries between highest uses, illustrated by points b, c, and d in Figure 15 which are the extensive margins of land use for particular purposes. Thus efficient land allocation constantly changes.

Finally, note that markets function only imperfectly in allocating land among uses. We have referred to alternative uses of land as being capable of generating land rent as if this were a continuous, or annual return to the land. Where markets operate effectively the rent becomes capitalized in the value of the land, in accordance with the relationship between annual returns and their present worth, described in Chapter 6. Competition among potential purchasers or renters ensures that each tract is allocated to the one who can pay the most for it, which must be someone who will put it to its highest use. However, this process often responds only slowly to changing economic conditions, and it is faulty whenever externalities in land use mean that the social benefits or costs associated with land use are not fully revealed in market prices.

COMBINATIONS OF USES

So far, we have considered the economics of producing various goods and services from a forest one at a time. We must now turn to the opportunities for multiple use, that is, the joint production of two or more goods or services.

Multiple use is a popular idea, frequently extolled as a means of reconciling the growing and often conflicting demands on natural resources. But it is also a vague concept, presenting a good deal of difficulty to resource managers who seek to apply it. When, and to what extent, is it *technically feasible* to accommodate two or more uses? In those cases where it is technically possible, when is it *desirable* to do so on economic or social grounds? And when multiple use is desirable *how much* of one use should be sacrificed for another? The remainder of this chapter examines these questions.

Interdependence and Production Possibilities

As already noted, the pattern of demands on land varies widely and so does the capability of land to produce various goods and services. The most productive use, or combination of uses, must be considered separately for each area in light of its particular circumstances. However, it is useful to identify the range of conditions with respect to the possibilities for multiple use.

At one extreme, land may be subject to no demand for any use. Some remote and inaccessible forest lands fall into this category. In Figure 15, they lie to the right of point e, beyond the extensive margin of forestry where rent-earning capacity declines to zero. These lands may contribute in a general way to the natural environment, but if they are not perceived as having values in any specific use the decision-making problem does not arise. They may, of course, take on some value in the future, but it is not until then or at least until a future value is anticipated, that decisions must be made about their allocation.

A second category consists of lands that are demanded for only one form of use. Many productive forests, rangelands, and remote recreational resources fall under this heading. They are illustrated in Figure 15 by the range of land over which only forestry is capable of generating a positive rent. This case raises the issue of the appropriate pattern and intensity of development, but there is no problem of allocation among competing uses.

It is worth noting that the absence of competing demands on land does not mean that it is technically incapable of other forms of production; it only means that it cannot generate an economic rent in other uses. The physical or biological capability of land to produce various goods and services presents an allocation problem only when two or more of them can be produced at a cost less than their benefits.

The remaining categories refer to lands that are capable of generating rent under two or more forms of production, raising the problem of choosing the best use or combination of uses. The efficient choice depends heavily on the technical interdependence of the uses, which can take several forms. Most common are *competing uses*, where the production of one product requires some compromise in the other. For example, forestland used for timber production can also be used for recreation, but recreational capacity can usually be expanded only by sacrificing some timber production, and vice versa.

This is like the well-known trade-off between "guns and butter" used to illustrate production possibilities in economics textbooks: if all factors of production in an economy were devoted to producing guns or butter, more of either of these products could be produced only by producing less of the other. The versatility of factors of production effectively enables one product to be transformed into another, according to a transformation or *production possibilities curve* like that in Figure 16a. The curve joins up all the possible combinations of timber and recreation that can be produced on a fixed tract of forest with the same amount of labour and other variable inputs.

The production possibilities curve in Figure 16a shows that without any provision for recreation it is possible to produce OT timber, measured in cubic meters of wood per year. And without any timber production OR recreation days could be accommodated. The points along the curve between these extremes indicate the possible combinations of the two products that can be produced with the same inputs.

The nearly horizontal slope of the curve near its intersection with the vertical axis indicates that it is possible to produce some recreation with very little sacrifice in timber production, but the more recreation produced the greater the required sacrifice in timber to produce another unit of recreation. Similarly, the more timber produced, the greater the sacrifice in recreation required to produce another unit. This gives the curve its concave shape, reflecting an *increasing marginal rate of transformation* of one product for the other. The curvature of the production possibilities curve thus reflects the degree of competitiveness of the two outputs, which varies over the range of possible combinations.

Any point inside the curve indicates a possible combination of the two products, but it implies that the resources available are not being fully utilized, or are being used inefficiently, because there are points on the curve showing that more of both products can be produced with the given inputs. The curve is thus a frontier of production possibilities.

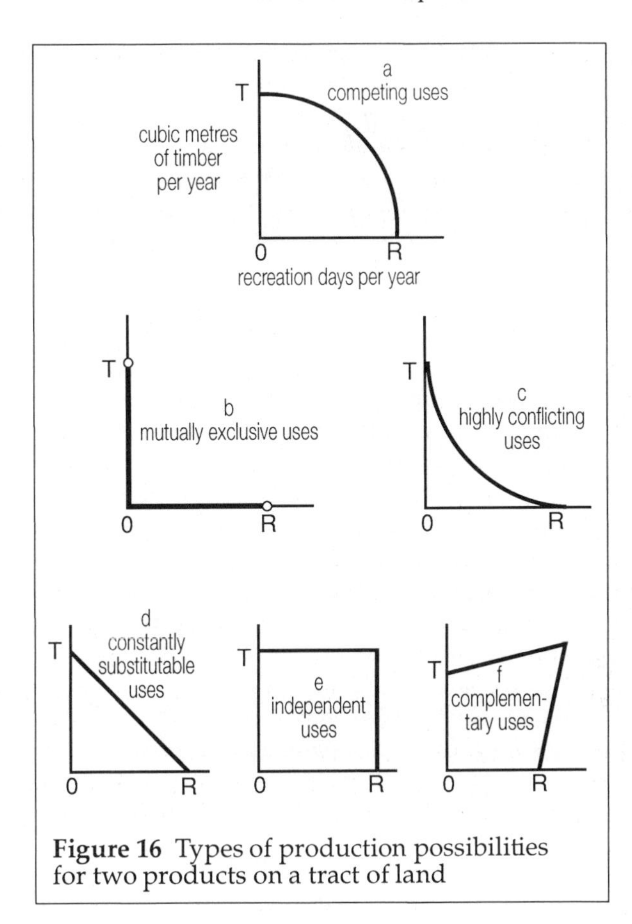

Figure 16 Types of production possibilities for two products on a tract of land

Production possibilities curves may take other forms, illustrated in the other quadrants in Figure 16. These are encountered less frequently in forest management, but they sometimes have important implications for land use decisions.

- *Mutually exclusive uses* are uses which are entirely incompatible. An example is timber production and preservation of a virgin forest for its scientific value. Thus Figure 16b shows the quantities of each of two products that can be produced, OT and OR, but no combinations of them.
- *Highly conflicting uses* are those where successive increments in the output of one product can be accommodated with progressively smaller sacrifices of the other. This relationship is unusual, but in some circumstances the trade-off between

timber production and amenity values can take this form, where a little industrial forestry would have a significant aesthetic impact on an otherwise undisturbed landscape, but successive increments of timber production would have smaller effects. This relationship is illustrated by the convex production possibilities curve in Figure 16c, which implies a *decreasing marginal rate of transformation* of one output for the other.

- *Constantly substitutable uses* are those for which the trade-off between the two products remains the same throughout the full range of production possibilities. Two such products produced on a tract of forest may be fuel wood and industrial timber, though there are few examples in forestry of such a *constant marginal rate of transformation*. This case is illustrated by the straight line production possibilities curve in Figure 16d.
- *Independent uses* have no affect on each other. Managing a forest for purposes of watershed protection may have no impact on recreational values, for example. Independent production possibilities are illustrated by the right-angled curve in Figure 16e which indicates that either product can be produced without impinging on the other.
- *Complementary uses* are found where one form of production enhances another. For example, management of a forest for timber production may benefit wildlife or range values in some circumstances. The production possibilities curve in Figure 16f thus illustrates how the production of one product increases the capacity to produce the other.

These two-dimensional diagrams illustrate the relationship between only two products. To illustrate similar relationships among three products it would be necessary to add a third axis, at right angles to the other two, and the production possibilities curves would take the form of three-dimensional curved surfaces and planes.

The impact of a marginal change in the production of one product on the capacity to produce a joint product varies with the intensity of land use. Most importantly, non-conflicting uses are frequently found at low intensities of use, while highly intensive management of land for any purpose often gives rise to conflict with other values.

Relative Values and Optimum Combinations

Even if two products or services can be produced on a tract of land and generate rent it is not always advantageous to produce both; to accommodate a second use may impinge so heavily on the first that the aggregate net value generated may be reduced. Moreover, even when two or more uses can be beneficially served, there remains the question of how much one ought to be sacrificed for another, that is, the appropriate compromise between the two.

As long as the objective is to generate the maximum land rent, account must be taken of the relative values of the interdependent products. Graphically, this can be represented by the slope of an *exchange line*, such as V_tV_r in Figure 17, where the amount of timber OV_t is equal in value to OV_r recreation. What is important here is the *relative* value of the two products, which governs the slope of the exchange line; in the figure this line is positioned so that it is just tangent to the production possibilities curve, at the point E.

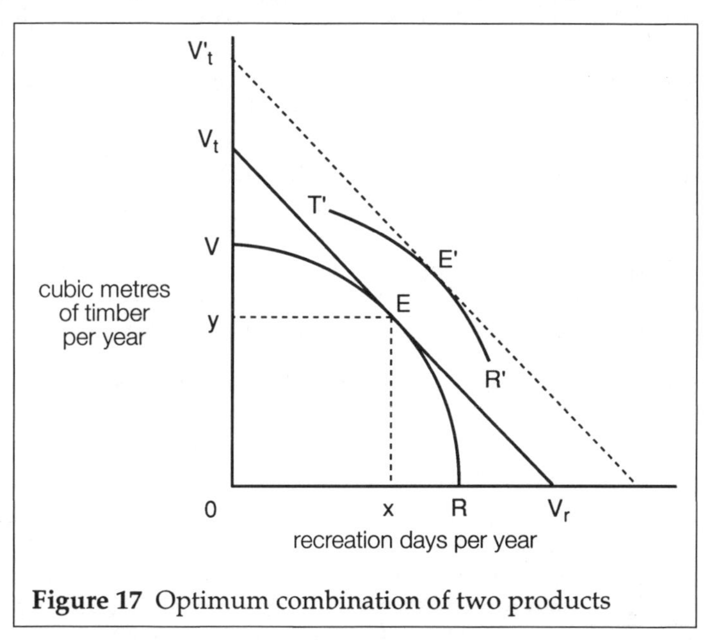

Figure 17 Optimum combination of two products

This point, E, indicates the optimum combination of the two products—Oy timber and Ox recreation—since no other possible combination will yield as high a total value. To the left of E, the more gradual slope of the production possibilities curve than the exchange line means that additional recreation is worth more than the necessary sacrifice in timber. To the right of E, total value can be increased by sacrificing recreation for timber. Thus, by shifting production in favour of one product as long as it is worth more than the value sacrificed in terms of the other product, the best combination is found at the point where the trade-off in physical possibilities is just equal to the trade-off in value between the two products, that is, where the marginal rate of transformation of one product for the other is just equal to the ratio of their marginal values. The highest total value of combined production is therefore at the point where the two curves are parallel, at the combination indicated by E in Figure 17.

With other forms of production possibilities, illustrated in Figure 16, the corresponding solutions are more straightforward. Independent and complementary uses should always be accommodated if they individually yield a net return, because as long as they do not impinge on other values this will increase the aggregate rent to the land. The point at which a sloping exchange line would be tangent to the production possibilities curves in Figures 16e and 16f is sharply defined and indicates combined production of both products. For mutually exclusive, highly conflicting, and constantly substitutable uses the solution will always be to produce only one product, whichever can generate the greatest rent. Thus, in Figures 16b, 16c, and 16d, a sloping exchange line would intersect the production possibilities curve where it intersects either the vertical or horizontal axis, indicating that the maximum rent can be generated by producing only the product measured on that axis.

Expanded Possibilities with Additional Inputs

The preceding representation of joint production possibilities is based on an assumption of fixed inputs. If the inputs available for production are variable, the frontier of production possibilities is not constrained to a single curve as described above. With a bigger land base or more labour and capital, the production possibilities curve will shift outward, indicating that more of either or both products can be produced, as illustrated by the curve T′R′ in Figure 17.

On the expanded frontier the new optimum combination of products is E′. The value of this increased output of both products is indicated by $V_tV'_t$ multiplied by the price of timber. This expansion of production will be advantageous if the cost is exceeded by this additional value of outputs.

The relationships depicted in Figure 17 closely parallel those used in the Appendix to Chapter 2 to demonstrate efficient factor proportions in production and returns to scale.

PRACTICAL DIFFICULTIES

If market economies functioned perfectly, as described in Chapter 2, financial incentives would induce landowners to allocate land to its highest use. In spite of their imperfections, most western industrial economies rely heavily on private landowners to determine how land is to be used. But markets for forest land often cannot be relied upon to allocate resources efficiently. Forestland is often publicly owned, and thereby withheld from market processes. Externalities in land use result in zoning and other controls, taxes distort economic incentives, and unpriced costs and benefits bias land use decisions.

The limitations of markets as means of determining the best patterns of land use put increasing onus on analyses of efficient solutions as described in this chapter. These economic analyses must be incorporated into land use planning and integrated resource management which have developed rapidly in recent years, and are now routine techniques for practising foresters.

This discussion reveals two dimensions to the problem of optimizing land use where more than one output is involved. One is to determine the most advantageous combination of products or services to produce at a given level of management intensity, that is, the point on a production possibilities curve, as discussed above. The other is to determine the optimum intensity of use, or how much to expand the production possibilities. Both these problems require attention to the economic values involved as well as to the technical possibilities of production.

For a simple illustration, suppose a forest produces a continuing yield of 1200 cubic metres of timber annually, and accommodates 1000 recreation days. Managers estimate that without increasing expenditures they could produce an additional 800 cubic metres per year by eliminating half the recreation capacity. The value of a cubic metre of timber is $20, and that of a recreation day $25.

The potential gain in the value of timber produced, at ($800 \text{ m}^3 \times \$20/\text{m}^3$) $16,000, exceeds the loss in recreation value, of (500 recreation days $\times$ $25/day) $12,500. This difference implies that the original combination was to the right of E on the curve TR in Figure 17, so the proposed change is closer to the economic optimum.

Suppose that this new combination was precisely the optimum, and that resource planners estimated that with additional annual management expenditures of $15,000 they could increase the production of both products by 50 per cent, that is by 1000 cubic metres and 250 recreation days. Clearly the extra value generated (of ($1000 \times \$20) + (250 \times \$25) = \$26,250$) exceeds the cost so expansion of the production possibilities is advantageous.

The practical application of the theoretical solution to optimum land use described in this chapter is rarely as simple as this example. Although the theory is straightforward, the data required to establish the relationships involved raise many difficulties. Two types of data are needed: the technical interdependencies of the products that underlie production possibilities and the economic information about their relative values. Technical data about the capability of forest land to yield specific goods and services is often sketchy. However, the crucial information required for resolving multiple use problems is not how much of each product can be produced in isolation but rather their interdependencies; that is, how much a little more of one product will impinge on the other over the range of possibilities. Relatively little research has been directed to these technical relationships, and the findings cannot be readily transferred from one situation to another.

The extent to which more intensive management can broaden production possibilities is also likely to be more complicated than our diagrams suggest. Depending on the products being considered and the type of resources available, the technical capacity to increase each product is likely to differ, and with more intensive production the relationship between the two is likely to change. So the expansion of the production possibilities frontier can be expected to be asymmetric and irregular. These are major obstacles to analyses of multiple land use opportunities.

Corresponding difficulties surround the economic data. The problem calls for information about marginal costs and benefits of each product. The cost of producing most products and services usually can be estimated, but where two or more products are produced jointly it is often difficult to separately identify the costs to be ascribed to each. For example, if reforestation of a site increases both recreational values and timber values, the cost must be prorated between them so that the costs can be compared with the benefits for each.

Measurements of benefits are often difficult also. Market prices are often unreliable, and must be corrected for market distortions. Most problematical are the values of unpriced benefits such as outdoor recreation and amenity, discussed in Chapter 4.

Finally, multiple use may involve various patterns of land use, and they are sometimes quite complicated. In some cases it means two or more products or services produced simultaneously over all the forest; in others it involves different uses spatially separated and scattered through the forest; in others it means different uses at different times during the forest cycle. The basic theoretical framework outlined in this chapter applies to all cases, but the practical problems of measuring the interdependencies varies considerably.

An additional dimension is added to the analytical problem when the different products or services accrue at different points in time. For these cases we need techniques for comparing values that are expected to be realized at different future times. This is the subject of the next chapter.

REVIEW QUESTIONS

1 How does an increase in the wage rate affect the amount of labour that must be employed to generate maximum returns to a tract of forest land? How do diminishing returns to labour influence this effect?
2 Why can more intensive silviculture be justified on highly productive land than on land of low productivity?
3 Explain how an increase in the price of timber would lead to (a) more intensive cultivation of forest land, and (b) a shift in the extensive margin of timber production.
4 Some land that finds its highest use in timber production is less productive in this use than some land devoted to agriculture. Does this mean that the land devoted to agriculture would be better used in timber production?
5 A forest produces 2000 cubic metres of timber annually, valued at $50 per cubic metre, and livestock forage that supports 500 cattle, worth $100 per head. The livestock-carrying capacity could be increased by 25 per cent by reducing the forest cover, but this would reduce the annual harvest of timber by 300 cubic metres. Would such a change improve the efficiency of land use?
6 Why can't market forces always be relied on to ensure that land will be allocated to its highest use or combination of uses?

FURTHER READING

Bowes, Michael D., and John V. Krutilla. 1989. *Multiple-Use Management: The Economics of Public Forestlands*. Washington, D.C.: Resources for the Future. Chapter 3

Campbell, G.E. 1981. Conceptual and empirical multiproduct production models in forestry: a survey. In E. Gundermann and F.H. Kaiser (comps.). *Research Today for Tomorrow's Forests: Proceedings for Outdoor Recreation Economics*. Seventeenth IUFRO World Congress, Kyoto, Japan, 6-12 Sept. 1981. [Munich: IUFRO]. Pp. 29-45

Clawson, M. 1978. The concept of multiple use forestry. *Environmental Law* 8:281-308

Convery, F.J. 1977. Land and multiple use. In M. Clawson (ed.). *Research in Forest Economics and Forest Policy*. Papers Resulting from a Symposium on Research in Forest Economics and Forest Policy, Resources for the Future, Washington, D.C. Pp. 249-326

Gregory, G.R. 1955. An economic approach to multiple use. *Forest Science* 1(1):6-13

Hof, John G., Robert D. Lee, A. Allen Dyer, and Brian M. Kent. 1985. An analysis of joint costs in a managed forest ecosystem. *Journal of Environmental Economics and Management* 12(2):338-52

Hyde, William F. 1980. *Timber Supply, Land Allocation, and Economic Efficiency*. Baltimore: Johns Hopkins University Press for Resources for the Future. Chapter 4

Pearse, P.H. 1969. Toward a theory of multiple use: the case of recreation versus agriculture. *Natural Resources Journal* 9(4):561-75

Sedjo, Roger A. (ed.). 1983. *Governmental Intervention, Social Needs, and the Management of U.S. Forests.* Washington, D.C.: Resources for the Future. Part 1

CHAPTER SIX

Valuation over Time and Investment Criteria

Out in the Forest . . .

As the company's forester, Ian Olson, prepares the long-term plans that guide Peavey Forest Products Limited in its road development, harvesting, and silviculture activities. The planning process forces him continually to compare values across time. How does the value of a harvest today compare with the value of harvesting the same timber five years from now? How does the cost of a pre-commercial thinning compare with the higher value of the harvest a decade later? Is the cost of a road improvement greater or less than the resulting savings in trucking costs over the next fifteen years?

In making these comparisons Olson must take account of the fact that a dollar next year is worth less to the company than a dollar this year, and a dollar fifty years hence is worth a great deal less. To do this he applies a rate of interest, or discount rate, to reduce future values to their equivalent present values, so all values can be compared in consistent terms.

Each year, the company's annual budget provides a fixed allocation for silviculture spending on the company's forestland. The allocation is never enough to undertake all the worthwhile silvicultural projects Olson has identified, so he must establish priorities. This he does by estimating the cost of each project and the equivalent present value of the future benefits that will result from it, choosing the projects that show the highest per dollar of benefits to costs. This assures him the maximum expected return from his limited budget.

Many forest management decisions involve choices about timing. At what age should the crop be thinned or fertilized? How fast should the forest inventory be harvested? How long should the crop be grown? Forest management involves planning a sequence of costs and benefits spread over time. This chapter explains how we can compare values that accrue at different times, to assist in assessing the courses of action available to forest managers.

Forests present almost infinite opportunities in silviculture, protection, access development, and other measures that can yield benefits. So we need, in addition to methods for evaluating actions, criteria for identifying those that justify the cost and those that should be given priority over others when choices must be made among them. Techniques for evaluating and comparing forestry investment opportunities are therefore examined in this chapter as well.

TIME AND THE ROLE OF INTEREST

Interest rates indicate how much more society values a dollar today than a dollar tomorrow. The *rate of interest* is therefore the key to comparing values that accrue at different times.

There are two explanations for putting more weight on present values than on future values. First, capital has an opportunity cost. Capital, like most other productive resources, can generate returns in alternative uses. When capital is tied up for a particular purpose, the sacrifice in other production must be taken into account. This *opportunity cost of capital* over time is measured by interest. Thus, if a million dollars worth of timber is held from one year to another, for example, the opportunity cost of doing so is the interest which the owner could earn if he liquidated the timber and invested the million dollars wherever it would yield the highest return.

The second explanation is the phenomenon of *time preference*. People generally prefer something today to something tomorrow. Savings behaviour demonstrates this; saving means postponing consumption, and savers demand some compensation for doing so. Interest is the reward for deferring consumption, or the return on saving.

In addition to measuring the effect of time on values there are two other uses of interest. One is to account for risk and the chance of failure. Investors must be compensated for taking risks, and differences in market rates of interest reflect varying degrees of riskiness in economic ventures. The riskier the venture, the higher the interest rate investors demand and therefore the higher the reward if it succeeds.

The other use is to correct for changes in the value of money that result from inflation and deflation. For example, if inflation erodes the value of money by 4 per cent per year, an investor who demands a *real* rate of return of 8 per cent will have to find investment opportunities that yield a *nominal* rate of return of at least 12 per cent. Real rates of return thus can be calculated by subtracting from the nominal rates of return a percentage equal to the inflation rate.

The market rate of interest is the price of investment capital, which rations the available supply of capital among those who demand it. The supply of capital funds is provided by savers whose time preference (the degree to which they prefer present over future consumption) is less than the interest they can earn on their savings. The demand for capital arises from investors wanting funds to invest in projects that will yield rates of return in excess of this price, or cost of capital. Interest therefore reflects, simultaneously, the return on investment and the reward for saving. The supply and demand for capital take the usual form; suppliers offer more when the price is high and demanders want more when the price is low, and the equilibrium price is the interest rate at which supply and demand are equated.

If the market for capital works effectively, all projects capable of yielding a return in excess of the market rate of interest will be taken up and projects with lower rates of return will not. Thus interest ensures that capital is allocated to its most productive uses. Moreover, as long as the supply is balanced by the demand for investment capital, interest serves to allocate resources efficiently over time because it reflects the tradeoff savers are willing to make between present and future returns.

For example, if someone demands a rate of interest of 8 per cent, he is indicating this tradeoff: he is willing to give up $1.00 today for $1.08 one year from now. In other words, $1.08 has a *present value* of $1.00. At a rate of interest, or *discount rate*, of 8 per cent, $1.00 today and $1.08 one year hence are of equal value. Interest thus enables us to bridge time in comparing values. The appropriate rate of interest to use is discussed later in this chapter; first we turn to the linkage between present and future values in more detail.

COMPOUNDING AND DISCOUNTING

As noted in the above example, interest rates enable us to compare a value in the present with a value expected to accrue in the future. The present value can be *compounded*, at the appropriate rate of interest, to indicate its equivalent value at the time the future value will occur. Conversely, the future value can be *discounted*, using the rate of interest, to obtain its equivalent present value. Either way, the two values can be compared at the same point in time. Compounding involves increasing a present value to its equivalent worth at a future time, while discounting is the reverse.

Compounding. To take the simplest case first, we often want to know how much an amount invested today will yield at some future date. If \$1.00 is invested at 8 per cent interest for one year it will then amount to \$1.08. If it is invested for two years, the \$1.08 accruing at the end of the first year will grow by another 8 per cent by the end of the second year, that is, \$1.08 (1 + .08) = \$1 $(1 + .08)^2$ = \$1.17. A three-year investment would add another 8 per cent to this total \$1.17 (1 + .08) = \$1 $(1.08)^3$ + \$1.26. This process is called *compounding* because after the first year, interest is earned (compounded) on not only the initial amount invested (the *principal*) but also on the interest earned in previous years.

The general formula for these calculations is

$$V_n = V_o(1 + i)^n \tag{1}$$

where V_n is the value to which an initial amount V_o will grow when invested for n years at an interest rate of i. For example, \$100 invested at 8 per cent for ten years has a future value of V_n = \$100 $(1.08)^{10}$ = \$215.90.

Compounding is thus a straightforward mathematical calculation which can be carried out readily with the help of a hand calculator or a compound interest table, such as that presented in the appendix to this chapter, which gives the value of $(1 + i)^n$ for the relevant i and n. The table shows, for example, that to compound a value over ten years at 8 per cent it must be multiplied by 2.159. Thus \$100 invested for this period at this rate of interest will grow to \$215.90, as in the example above.

Compounding can be used to compare values accruing at different times. We have already compared \$1.00 today with \$1.08 a year hence. For another illustration, suppose someone offered you either \$100 today or \$200 ten years from now. Which would you choose? If you knew you could invest money at 8 per cent, you would choose the \$100 today because, as we have already seen, over ten years it is capable of growing at 8 per cent to \$215.90.

Discounting. The reverse problem is to determine the present equivalent value of a future payment. For this purpose, we can simply transpose Equation 1 to

$$V_o = \frac{V_n}{(1+i)^n} \tag{2}$$

which expresses the present worth, V_o, of an amount V_n, to be received n years hence. Thus, again with reference to the discount factors in the appendix, we can easily calculate that the present worth of $1.00 received one year hence, at a discount rate of 8 per cent, is $V_0 = \$1.00 \div (1.08)^1 = \$.93$. If the $1.00 were to be received two years hence its present worth is $.86, and if it were ten years hence its present worth is only $.46.

This process of *discounting* future values is another means of comparing present with future values. In an earlier example, we compounded a present value of $100 to compare it with $200 ten years hence. Alternatively, we can compare these values by discounting the one receivable in the future to obtain its present worth. Thus, the present value of $200 ten years in the future is $\$200 \div (1.08)^{10} =$ $92.64, which is less than the alternative of $100 receivable today in that example.

PRESENT VALUES

Compounding and discounting enable us to measure and compare, in consistent terms, values that accrue at different times. To do this, we must represent the values not only in the same terms—dollars—but also at their equivalent values at the same point in time. So, for example, to evaluate a particular plan for a forest stand we may have to take account of costs of establishing the crop immediately, management costs that will be incurred annually or periodically, revenues from thinnings sometime during the growing cycle, and revenues from the final harvests. In order to compare the revenues with the costs to see if the plan for the stand is advantageous, or better than an alternative plan, we must reduce all the values to the common denominator of dollars at the same point in time.

Compounding and discounting allow us to choose any point in time to evaluate such a plan. For some purposes it may be convenient to evaluate all revenues and costs at the time the forest is to be harvested, or alternatively, all values can be reduced to their equivalent value at the date the forest is to be established. However, the point in time most commonly chosen for evaluating forest management plans is simply the present.

The calculation of *present value* is usually divided into a calculation of the present value of the revenues or benefits of the project and the present value of the costs of the project. The excess of the benefits, B, over the costs, C, is the *net present value*, V_0, of the project.

$$V_0 = B - C \tag{3}$$

The following paragraphs review the most common forms of present value problems in forestry and how they are calculated.

Present Value of Future Revenues and Costs

A common problem is whether it is advantageous to grow a crop of timber on a tract of vacant land. If it would cost an amount, C_p, to plant the crop at the outset, and the final harvest is expected to generate a net return of V_n in n years' time, the formula for calculating the net present value of this investment is obtained by combining Equations 2 and 3, that is,

$$V_o = \frac{V_n}{(1+i)^n} - C_p$$

So if the planting cost (C_p) were $1,000, the growth period (n) 40 years, the value of the harvest (V_n) $100,000 and the discount rate (i) 6 per cent, the net present value of the venture is:

$$V_o = \frac{100{,}000}{(1.06)^{40}} - 1000$$

$$= \$8{,}718$$

This indicates a net gain of $8718 created at a cost of $1000, or nearly $9.00 per $1.00 of cost, all these values being measured in terms of their present value equivalents.

Present Value of a Perpetual Annuity

Sometimes an asset like a farm or a forest will yield a regular annual return in perpetuity, and costs of management are often expected to be a recurring annual amount as well. We therefore want to calculate the present worth equivalent of a value that occurs each year into the future.

The formula for calculating the annual return, a, on an amount, V_o, invested at i per cent, is simply

$$a = i(V_o) \tag{4}$$

Transposed and solved for V_o, this formula becomes

$$V_o = \frac{a}{i} \tag{5}$$

which gives the present worth, V_o, of an amount, a, receivable each year in perpetuity at a discount rate of i. This formula can also be obtained from Equation 2 above, by adding together the present worth of a geometric series of equal amounts, a, receivable each year in perpetuity. That is,

$$V_o = \frac{a}{1+i} + \frac{a}{(1+i)^2} + \frac{a}{(1+i)^3} + \ldots + \frac{a}{(1+i)^\infty}$$

which can be reduced to Equation 5.[1]

Equation 5 enables us to calculate, as a lump sum present worth equivalent, the value of an amount that will occur annually forever. For example, if a tract of timberland is expected to yield an annual harvest worth $1000, the present value of this recurring revenue, at an interest rate of 8 per cent, is 1000 ÷ .08 = $12,500.

Present Value of a Finite Annuity

If the future series of annual payments is limited to a finite number of years, say n, we need a formula for a geometric series with n terms:

$$V_o = \frac{a}{1+i} + \frac{a}{(1+i)^2} + \ldots + \frac{a}{(1+i)^n}$$

This can be simplified to:[2]

$$V_o = \frac{a\,[(1+i)^n - 1]}{i\,(1+i)^n} \qquad (6)$$

Equation 6 is useful for calculating the present worth of a series of future costs or revenues that will occur each year for a certain number of years. For example, the present worth of a forest that will yield a harvest worth $1000 each year for six years, at 8 per cent interest, is

$$V_o = \frac{1000\,[(1.08)^6 - 1]}{.08(1.08)^6}$$

$$= \$4,622$$

Present Value of a Periodic Series

Of the considerable variety of other formulae for special kinds of problems, the one most commonly needed for forestry purposes is that for determining the present worth of a future series of revenues or costs that will accrue only periodically at regular intervals of several years. If the interval is t years, and the net revenue received every t years is V_t, then

$$V_o = \frac{V_t}{(1+i)} + \frac{V_t}{(1+i)^{2t}} + \frac{V_t}{(1+i)^{3t}} + \ldots + \frac{V_t}{(1+i)^{\infty}}$$

This simplifies to:[3]

$$V_o = \frac{V_t}{(1+i)^{t}-1} \qquad (7)$$

This is the formula required to calculate, for example, the present worth V_o, of a future infinite series of forest crops having a harvest value V_t, which will accrue after each crop rotation period, t. Thus, if a tract of forestland can produce a crop worth $100,000 in forty years and every forty years thereafter, and the interest rate is 8 per cent, the present worth is:

$$V_o = \frac{100,000}{(1.08)^{40}-1}$$

$$= \$4,826$$

Incidentally, the low present worth of these crops results from the force of discounting at 8 per cent over forty years and more. Experimentation with this result and Equation 2 reveals that the second and subsequent crops contribute little to this value.

Now, to incorporate preceding formulae, let us elaborate this example by assuming two realistic complications. First, it will cost an amount, m, every year to manage and administer the forest. Second, the forest must be planted at a cost, C_p, at the outset and at the time of each harvest. Then the solution takes the form

$$V_o = \frac{V_t - C_p}{(1+i)^t-1} - \frac{m}{i} - C_p$$

The first term on the right side of the equation is the present worth of the series of future crops net of the cost of reforestation. The second term allows for the recurring annual costs of management, and the third accounts for the initial cost of planting the first crop. If, for example, harvests are expected to yield (V_t) $100,000, planting costs (C_p) are $1000, annual management costs (a) are $100, the discount rate (i) is 8 per cent, and the rotation period (t) is forty years, the net present value of this program is:

$$V_o = \frac{100{,}000 - 1000}{(1.08)^{40} - 1} - \frac{100}{.08} - 1000$$

$$= \$2{,}528$$

A variety of other complications in costs and returns can be incorporated into such problems.

The above formulae for compounding and discounting, and some others commonly used in forestry, are summarized in Appendix B to this chapter. For simplicity throughout this book, all formulae and calculations of values over time are based on annual compounding and discounting. It should be recognized, however, that values can be compounded or discounted with any periodicity, which in practice is often semi-annually or even more frequently. The more frequent, the greater the force of interest. At the extreme, compounding can be continuous, which would cause \$100 to grow to \$112.75 in one year at 12 per cent compared to \$112.00 under annual discounting.

CRITERIA FOR INVESTMENT DECISIONS

To evaluate opportunities for forestry investments we again concentrate on the criterion of efficiency—the benefits relative to the costs. The benefits and costs occur at different times; indeed, it is a characteristic of all investment opportunities that they involve an initial cost which gives rise to future benefits and perhaps additional costs spread over time. Evaluation involves weighing the benefits against the costs, taking account of the timing of their occurrence, to establish their relative efficiency. The most efficient management plan will generate the greatest possible benefits for the costs incurred.

The technique for weighing benefits and costs is called *benefit-cost analysis*. It can take several forms, each appropriate for a different purpose. For the time being, we will maintain the simplifying assumption that all the costs and benefits we have to consider are known with certainty.

Identifying Advantageous Investments

In considering an array of possible forestry investments the first task is often to sort out those that are advantageous from those that are not. Advantageous investments are those that yield benefits in excess of the costs, or a *net benefit*. More specifically, whenever the relevant values occur at different times, the net benefit is the difference between the present value of benefits and the present value of costs. Where a project indicates a net benefit greater than zero it is economically *feasible* in the sense that it will generate returns in excess of all costs, including the opportunity cost of the capital involved. Where the expected benefits fall short of the costs, the project is economically unfeasible in the sense that society would be made worse off if it were undertaken. In other words, the value of the outputs must exceed the value of the inputs, otherwise the inputs could generate greater value in producing other things. This provides a basic rule for identifying investment opportunities that are beneficial: the benefits must exceed the costs, both expressed in present value. In terms of Equation 3:

$$V_o = B - C > 0.$$

The indication of a positive net benefit is usually considered a necessary condition for worthwhile investments but, as we shall see, it is rarely a sufficient justification for undertaking them.

Identifying Priorities

Which of all the economically feasible forestry projects should be undertaken? The answer depends on the circumstances. If all the projects were independent of each other, and there was no limit to the capital and other resources available, then they all might be undertaken because all will generate a net gain. But the available resources are more often limited, or one project is a substitute for another, or they are alternative ways of using the same site, so we need criteria for selecting among projects that show varying degrees of economic feasibility.

Net benefit. The greater the surplus of benefits over costs, the greater the gain from a project. Alternative feasible investment opportunities can be compared and ranked according to the magnitude of the net benefit they are expected to generate, and those ranking highest given highest priority.

Where there are alternative ways of using some fixed resource such as a tract of land, and the task is to find the one that will generate the highest returns, net benefits provides the appropriate guide. For example, a forest owner might have to decide which of three mutually exclusive uses of a forest tract is most advantageous: timber production, agriculture, or wilderness preservation. The net benefits of each can be compared and the one promising the greatest net benefit selected because it, by definition, would generate the greatest gain.

Consider a typical forestry calculation of this kind, where a tract of forty hectares has been harvested and the choice is to plant it or leave it to regenerate naturally. If planted immediately, at a cost of $400 per hectare, the crop harvestable in fifty-five years would yield 475 cubic metres per hectare worth $25 per cubic metre. Alternatively, if left to reforest itself naturally at no cost, the crop in fifty-five years would yield only 375 cubic metres per hectare of an inferior species having a value of $15 per cubic metre. The problem is to identify which alternative, planting or natural regeneration, is most advantageous or efficient.

The benefit of the investment in planting is the difference between the present net value of the forest when planted, V_p, and its present net value when left to regenerate naturally, V_g. The cost, C, of planting the forty hectares is $16,000, which needs no discounting because it would be expended immediately. Thus, using a discount rate of 3 per cent, the net benefit is

$$B - C = (V_p - V_g) - C$$

$$= \frac{40 \times 475 \times 25}{(1.03)^{55}} - \frac{40 \times 375 \times 15}{(1.03)^{55}} - 12{,}000$$

$$= 49{,}194 - 16{,}000$$

$$= \$33{,}194.$$

This positive net benefit indicates the advantage of planting over the alternative of not planting on this site. Such calculations thus identify the economically most advantageous silvicultural regimes from the range of alternatives.

Net benefit is the appropriate basis for choosing among mutually exclusive uses of some fixed resource where there are no artificial limits on other inputs such as investment capital. In the next chapter we examine how corresponding criteria are used to select the most advantageous harvest age for a forest stand, and how the most efficient management regime determines the land rent.

Benefit/cost ratio. Forest managers seldom find themselves with the unlimited capital and other resources to take advantage of all projects that promise to yield positive net benefits. When funds are limited, and some feasible projects must be selected over others, the problem becomes one of generating the maximum possible benefits from the funds available. This means, in other words, generating the maximum possible benefits per dollar invested, or maximum beneits relative to cost. A project's total benefits relative to its total cost indicates its *benefit/cost ratio*. This ratio corresponds to the criterion of efficiency described in Chapter 1; by expressing outputs relative to inputs it measures the efficiency with which resources are used.

The earlier example of a potential investment in planting a new forest indicated benefits, B, of $49,194 and costs, C, of $16,000. This yields a benefit-cost ratio of

$$\frac{B}{C} = \frac{49,194}{16,000}$$

$$= 3.07$$

While planting was shown to generate more net benefits from this forest tract, the investor may have to weigh this use of funds against some competing project. Suppose, for example, he could alternatively space a juvenile stand of sixty hectares, at a cost of $550 per hectare, which would then yield a harvest in fifty years of 550 cubic metres per hectare worth $25 per cubic metre. Without this treatment, the harvest in fifty years would be only 425 cubic metres per hectare, worth $15 per cubic metre.

The benefit of the investment in spacing is the extent to which the present net value of the spaced stand, V_s, exceeds the present net value of the unspaced stand, V_u. The cost of spacing the sixty hectares is $33,000. So the net benefit of the investment is

$$B - C = (V_s - V_u) - C$$

$$= \frac{60 \times 550 \times 25}{(1.03)^{50}} - \frac{60 \times 425 \times 15}{(1.03)^{50}} - 33,000$$

$$= 100,937 - 33,000$$

$$= \$67,937.$$

This project is therefore economically feasible as well; indeed, it indicates a larger net benefit than the planting project. But note that this project shows a benefit/cost ratio of

$$\frac{B}{C} = \frac{100,937}{33,000}$$

$$= 3.06$$

which is slightly lower than the corresponding ratio for the planting project. It is therefore a less efficient investment because it generates a lower return per dollar invested.

The benefit-cost ratio provides a criterion for selecting among a variety of independent projects in order to generate the maximum return to the available budget. If projects are given priority based on their benefit-cost ratio, and the available funds allocated to them accordingly, the returns *per dollar* invested will be maximized and, as a result, the maximum possible benefits and net benefits will be generated from the limited investment funds. However, the ranking of projects according to their benefit/cost ratios will rarely correspond to the ranking according to their net benefits.

Return on investment. A third method of assessing investment projects involves calculating their *internal rate of return*. Instead of comparing the present value of benefits and costs discounted at some predetermined interest rate, this technique involves finding the interest rate, or discount rate, that equates the present value of benefits with the present value of costs. The higher the indicated rate of return on the investment, the more attractive is the project.

The internal rate of return of a project is thus the percentage rate at which the initial investment grows over the investment period to the value of the benefits. Unlike the preceding criteria, which relate benefits to the cost of all inputs, including capital, this technique treats the return on capital as the residual, after all other costs are accounted for.

In our first numerical example of an investment in planting above, an expenditure of \$16,000 would yield a gain, in fifty-five years, equal to the value of a planted stand in excess of the value of a naturally regenerated stand, that is,

$$(40 \times 475 \times 25) - (40 \times 375 \times 15) = \$250,000.$$

The internal rate of return is the rate, r, at which \$16,000 grows over fifty-five years to \$250,000, that is,

$$16,000\ (1 + r)^{55} = 250,000$$

$$r = 5.12\%.$$

The second example involved an initial spacing cost of \$33,000 for a gain in fifty years of additional stand value equal to

$$(60 \times 550 \times 25) - (60 \times 425 \times 15) = \$442,500,$$

which indicates an internal rate of return of

$$33,000\ (1 + r)^{50} = 442,500$$

$$r = 5.36\%.$$

The internal rate of return criterion for assessing investments is used mainly by investors of capital seeking to maximize the return on their money. It eliminates the need to select a rate of interest in advance, though sometimes investors consider only investment opportunities that indicate a rate of return in excess of some minimum acceptable rate. The internal rate of return has the additional appeal of simplicity; most people understand that a higher rate of return on an investment is preferable to a lower one.

However, this criterion presents serious problems in application to many projects and can give misleading results. It is suitable only for projects that involve an initial investment and later returns. Some projects generate early returns and involve later costs, which appear more favourable the higher the discount rate. Such projects can be justified only if the internal rate of return is *less than* the investor's maximum rate. Moreover, calculation of the internal rate of return becomes exceedingly complicated, and sometimes indeterminate, when a project involves positive and negative returns scattered through time. Finally, because it fails to take explicit account of the opportunity cost of capital this investment criterion cannot indicate which investments are desirable from the viewpoint of society as a whole.

Comparing Criteria

To illustrate the relationship among these investment criteria let us consider a third numerical example. A 100-hectare forest scheduled to be harvested in fifteen years would normally yield 350 cubic metres per hectare worth $15 per cubic metre. However, an insect infestation threatens to destroy 25 per cent of the timber unless control measures costing $350 per hectare are undertaken immediately.

The benefit of insect control is the present worth of the 25 per cent of the stand that would be saved and realized in fifteen years. The cost of spraying is $35,000. So the net benefit, at a discount rate of 3 per cent, is

$$\begin{aligned} B - C &= \frac{100 \times 350 \times 15 \times .25}{(1.03)^{15}} - 35,000 \\ &= 84,244 - 35,000 \\ &= \$49,244. \end{aligned}$$

The benefit-cost ratio for this project is

$$\frac{B}{C} = \frac{84,244}{35,000}$$

$$= 2.41$$

and the internal rate of return is

$$35,000\ (1 + r)^{15} = 100 \times 350 \times 15 \times .25$$

$$r = 9.21\%.$$

Table 1 summarizes the results of applying each of the three evaluation criteria to each of the three examples above.

TABLE 1: Comparison of investment projects using alternative evaluation criteria

Criterion	Planting project	Spacing project	Pest control project
Net benefit (B – C)	$33,194	$67,937	$49,244
Benefit/cost ration (B ÷ C)	3.07	3.06	2.41
Internal rate of return (r)	5.12%	5.36%	9.21%

All three projects are economically feasible insofar as they generate returns exceeding their costs. Thus the net benefit is positive in each case, and the benefit-cost ratio is greater than 1.0. Obviously, modifications to any of the examples, such as higher costs, lower or later benefits, or higher discount rates would reduce the net benefit and, if sufficient, reduce it to a negative value, rendering the project unfeasible.

More importantly, the *relative* attractiveness of the projects differs under different evaluation criteria. Net benefits are greatest for the spacing project, yet this project does not show the highest internal rate of return. The planting project has the highest benefit-cost ratio but the lowest net benefit.

Because these criteria often suggest different investment priorities, it is important to understand the appropriate use of each. To begin, it should be noted that the theoretical conditions for a perfect market economy, outlined in Chapter 2, imply that resources are deployed in such a way that they all earn their opportunity cost at the margin in all uses. This implies, in turn, that the most attractive new investments are marginal, showing no benefits in excess of costs, benefit-cost ratios in excess of 1.0, or rates of return exceeding the opportunity cost of capital. The problem of choosing among investment criteria arises from market imperfections that create opportunities for earnings in excess of opportunity costs.

The *benefit-cost ratio* corresponds to the criterion for efficiency described in Chapter 1. By expressing outputs relative to inputs it measures the efficiency with which resources are used. An individual decision-maker who, like society as a whole, is indifferent about particular resources and simply seeks to maximize the benefits for the resources expended, would adopt this criterion.

In contrast, a decision-maker looking for the way to generate the maximum returns to some fixed factor such as a tract of land should, as illustrated, choose the alternative that will generate the maximum *net benefit*. Where the investor seeks the maximum return on money capital, the criterion should be the *internal rate of return* on capital invested. The appropriate criterion thus depends on which factor of production, if any, is limited for the decision-maker, inducing him to maximize returns to it.

The indivisibility of investment projects sometimes necessitates modification of these rules. For example, if the decision-maker's budget was $12,000, he could undertake only the planting project regardless of the evaluation criterion. If the budget were triple that amount, he could choose any one of the three projects, but if he chose the planting project he would have $24,000 left over, too little to undertake either of the remaining projects. If, in this case, he seeks to maximize the returns on his budget (as opposed to project costs) and has no opportunity to earn a return on residual funds, he would find that both the other projects would show a higher benefit/cost ratio. Of course, if there were three potential projects like the planting project, allocating the budget to them would generate the highest benefit/cost ratio.

The net benefit criterion favours large projects. If the planting project were three times larger, the tripled net benefit would raise the ranking of this project from third to first. This would not change either the internal rate of return or the benefit-cost ratio, however. The internal rate of return criterion favours projects with low initial capital costs and early returns. Note that the pest control project, which yields benefits much sooner than the other projects, shows the highest internal rate of return but ranks lower by the other criteria.

Finally, it should be noted that the ranking of projects by either the net benefit or benefit-cost ratio criteria is sensitive to the interest rate. Higher interest rates favour projects that yield early returns and later costs. Changing the interest rate can change the relative attractiveness of projects by different criteria and by a single criterion.

INTEREST RATE

The rate of interest used in evaluating investments is thus critically important in determining the results. However, the appropriate rate to use is often problematical.

In principle, the appropriate rate of interest is the investor's opportunity cost of capital; that is, the rate that he must pay to borrow or the rate he can earn on capital invested elsewhere at the margin. This rule applies to both governmental and private investors. But financial markets reveal a spectrum of interest rates, reflecting varying allowances for risk, expected inflation, distortions of the tax system, other market imperfections, and real returns to capital. Market rates also fluctuate continuously. It is difficult to identify from this information the opportunity cost of capital to apply to a specific investment and investor.

A first approximation of the opportunity cost of capital free of the distortions of risk, short-term disturbances, and inflation is often made from the historical average yield on long-term government bonds. These securities, bought and sold in large quantities in competition with private securities, are considered virtually risk-free because they represent loans to governments, and governments are unlikely to default on their debts because of their control over tax revenues and the money supply. In Canada and the United States this rate, after adjusting for inflation (see below) has historically been in the order of 1 or 2 per cent. Real rates of return on private equity capital have been considerably higher, in the order of 6 per cent. However, these rates reflect much higher allowances for risk, they usually fail to account fully for negative returns on unsuccessful investments, and they measure realized returns that do not necessarily coincide with expectations at the time investments are made.

From the viewpoint of society as a whole, efficient allocation of capital requires that the returns on private and governmental investments are equal, apart from different allowances for risk. However, the returns on private projects are usually taxed while governmental earnings are not, so private rates of return must be higher to yield equal *after-tax* returns to private investors.

A more fundamental problem is whether the rate at which a society at large wants to give preference to present over future values, the *social rate of time preference*, bears any relationship at all to the rates that individual savers and investors, acting independently, generate in capital markets. Despite extensive economic and philosophical enquiry, the social rate of time preference remains elusive.

To the extent that it differs from the market rate of interest it must be decided collectively through political processes, but this is never done explicitly.

So the search for the appropriate social rate of interest can take only limited guidance from observed market rates. Nevertheless, the highest prevailing market rate of return, adjusted for risk, may provide a useful indication of the opportunity cost of capital. And because of the prevalence of imperfections in capital markets it may be regarded as an upper limit to the social rate of discount.

For operational purposes investors usually turn to more immediate indicators of their opportunity cost of capital, such as the rate at which they can borrow or the rate they can earn on other investments. On this basis both governments and private investors often specify a required rate of return to be used in evaluating their investment opportunities.

Because of the long investment periods in forestry, the results of evaluations are highly sensitive to the interest rate chosen. Forestry investments are often evaluated at relatively modest real rates of 2 or 3 per cent. Low rates are sometimes justified on grounds of low risk, but the riskiness of forest investments varies widely and must be considered in each case.

ALLOWING FOR INFLATION

Inflation is a rising general price level or, to put it another way, a declining value of money. It distorts dollar values over time. So values accruing at different points in time must be corrected for inflation before they can be compared on a consistent basis.

Over the long periods of time considered in forestry problems, even a modest rate of inflation can have a large impact on prices and costs. For example, annual inflation of 3 per cent, compounding over time like interest on capital, will double prices in less than twenty-five years. In this case, if an amount receivable twenty-five years hence, expressed in the *current dollars* of that time, is to be compared with values today, it must be reduced by more than half so that the values can be compared in *constant dollars*.

Values accruing in different years, expressed in their current dollar values, can be corrected for inflation by deflating them, using a *price index* which measures the change in the price level over the years. For example, with an inflation rate of 3 per cent, a price index based on a value of 100 in the current year will be 103 next year and 209 in twenty-five years. By dividing values that occur in different years by the price indices for the corresponding years, all values are reduced to constant dollar values. The best-known index of inflation is the consumer price index, based on the change in price of a representative sample of consumer goods and services. However, other indices, such as the wholesale price index, are usually more suitable for projects of an industrial nature.

Interest rates can also be adjusted for inflation. The rates of interest we observe in money markets embody two components: the *inflation rate* and the *real rate*. The inflation rate is simply an allowance for inflation; it is the rate at which a value must grow in current dollars simply to maintain its value in constant dollars. The real rate reflects the return to capital after inflation is allowed for, the nominal rate minus the inflation rate.

Projects can be evaluated in terms of the nominal costs and benefits expected at future times, embodying expected inflation. Then, discounting at a rate that includes the inflation rate will reduce inflated future values to their present real values.

However, because the rate of inflation is difficult to predict over long periods, and in any event is likely to change over the investment period, projects are more often evaluated in terms of *constant* dollars, abstracting from inflation. All costs and benefits are estimated in dollars having the same value, usually their value in the present year, or the year the project would be undertaken, and are reduced to their present values using only a *real* rate of interest.

RECOGNIZING UNCERTAINTY

So far we have discussed the revenues and costs associated with an investment project as if they are known, or can be predicted accurately. But future revenues and costs are always more or less uncertain, and the further into the future they are expected to occur the more uncertain they are.

Future costs and revenues associated with forestry projects are often highly uncertain, especially when they are based on predictions spanning several decades. Knowledge about how stands grow and respond to treatments is always limited. Expectations about future harvests can be upset by unpredictable events such as fire and other natural catastrophes. And the technology, product prices, and production costs assumed in making predictions are likely to change in unforeseeable ways.

Sometimes a distinction is made between risk and uncertainty on the basis that the former lends itself to prediction while the latter does not. For example, future prices of timber are uncertain because there is no statistical basis for predicting them. In contrast, the risk of forest fire can be statistically estimated, just as insurance companies estimate the probabilities of house fires. Moreover, risk can be spread and thereby eliminated.

For example, if a tract of forest is the only asset of a small landowner, the risk that it may be destroyed by fire, though small, is likely to be of major consequence to him. In contrast, if it is only one of thousands of such tracts held by a large landowner his average fire losses are likely to be statistically predictable, and so can be allowed for without bearing much risk. This is analogous to a house insurance company which assumes the risk of fire losses of thousands of individual houseowners, and by spreading the risk over large numbers effectively eliminates it. This means that although the inherent risks associated with a project must be accounted for regardless of the investor, his response will be influenced by his circumstances. And, generally, large diverse investors, especially governments, are likely to be less averse to risk-bearing to the extent that they can spread its effects.

Generally, investors are *risk averse*; that is, faced with two investment opportunities having equal expected returns but one being riskier than the other, most investors would prefer the less risky one. Accordingly, investors demand higher returns from risky ventures. Risk thus adds an extra cost to a project; the more uncertain its outcome, the greater the risk premium investors will demand on their expected returns.

There is no single accepted technique for allowing for uncertainty in analysing investments. However, several criteria have been developed to assist managers and investors in making consistent decisions in the face of uncertainty; the appropriate choice among them depends on the investor's objectives and his attitude toward risk-taking.

Risk Premiums

One way of taking account of uncertainty is to increase the interest rate used in the evaluation by a premium sufficient to compensate the investor for the riskiness of the project. The size of the risk premium that should be added to the risk-free interest rate depends on the inherent riskiness of the project and the investor's aversion to risk-taking.

Adding a risk premium to interest rates is a simple procedure, but it presumes that the uncertainty surrounding future revenues and costs is perfectly correlated with their time of expected occurrence. This is not usually the case. Some future costs and revenues, such as insurance premiums and property taxes, can be estimated quite closely for years into the future, while others, such as timber prices and fire-fighting costs, are often difficult to predict even over a relatively short time span. Applying a risk premium to the interest rate means discounting all of them by a factor that depends only on how far into the future they will occur, thus blurring their differences in uncertainty.

Payback Period

A simple rule used by some investors is that any project must be capable of generating returns sufficient to cover the cost of the investment within a certain period. Fixing a maximum payback period does not explicitly recognize the uncertainty of future costs and returns; it is simply a decision rule that recognizes the most likely outcome within the period, and nothing beyond it. The payback period used by private investors is often short, of five years or less, which rules out superior (and even more certain) projects if their returns are delayed for many years, as in most forestry investments.

Analysing the Range of Possible Outcomes

Another way to assist decision-makers considering uncertain projects is to evaluate them not only in terms of their most likely outcomes but also for the full range of other possible outcomes, from the most pessimistic or "worst case scenario" to the most optimistic. The results give investors a feel for the range of possible acceptable and unacceptable outcomes, and for their dispersion around the outcome considered most likely.

To provide additional guidance, the analyst may, for each possible outcome, provide an estimate of the probability that it will occur. Then each possible outcome can be weighted by its probability to portray in a more meaningful way the likelihood that the project will yield acceptable results. Assigning probabilities is often problematical, however; usually empirical information suitable for this purpose is scanty and estimates must be made subjectively. Nevertheless, the process forces analysts and decision-makers to be explicit, and hence more consistent, in recognizing the relative probability of different possible outcomes.

Once probabilities have been attached to the alternative possible outcomes, the decision-maker can, if he wants, use any one of a variety of criteria designed to respond to risk. One way of organizing a decision problem involving uncertainty is in the format of a so-called "decision tree," illustrated in Figure 18. This is a diagrammatic representation of the choices facing the decision-maker, the possible outcomes and the uncertainties associated with them, and the order in which events take place.

Figure 18 illustrates a decision tree for the pest control problem described earlier in this chapter, though we now introduce uncertainty. The branches from the square indicate the decision-maker's choices; branches from circles represent uncertain events. So the branches from the square on the left indicate that the decision-maker may choose to spray or do nothing. The upper branches indicate the possible outcomes if the spraying is undertaken: the expected infestation may not actually occur, in which case the control measures would be wasted, or it might occur, and then the spraying may either fail or succeed. Correspondingly, the lower branches indicate that if the decision-maker chooses not to spray the expected infestation may or may not occur.

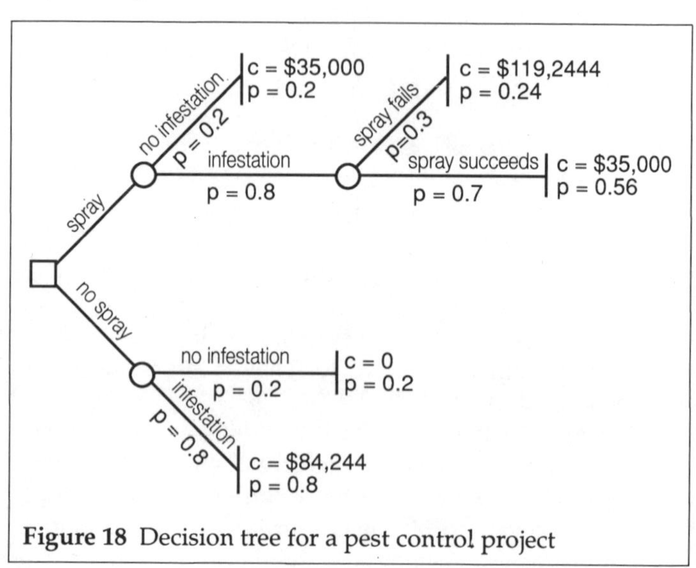

Figure 18 Decision tree for a pest control project

The total cost or loss, C, of each possible outcome is shown at the end-points of the decision tree. If the forest is sprayed and no infestation occurs, the loss is the cost of the spraying, $35,000 as shown, and this same cost would be incurred if the infestation occurs but the spraying is successful. If it fails, however, the total cost will be the sum of the cost of spraying and the loss of timber value calculated earlier, that is, $35,000 + $84,244 = $119,244. If no spraying is done and infestation occurs the cost will be the loss in timber value alone, \$84,244.

The probability, p, of each uncertain event occurring is noted along each branch, the sum of the probabilities for the possible events at each stage adding to 1.0. The probability of each final outcome, also indicated at the end-points of the tree, is the product of the probabilities of the events leading to it.

To assist in the decision about whether to spray or not to spray in this example we can calculate the expected cost of each alternative by weighting the cost associated with its various possible outcomes by their corresponding probabilities and adding them. Thus the expected cost of spraying is

$$.2(35{,}000) + .24\ (119{,}224) + .56(35{,}000) = \$55{,}219$$

and the expected cost of not spraying is

$$.2(0) + .8\ (84{,}244) = \$67{,}395.$$

The expected cost of not spraying exceeds the expected cost of spraying, so the criterion of minimizing expected losses (or maximizing expected values) suggests that the preferred choice is to spray.

Other criteria or decision rules can be adopted in recognition of decision-makers' differing attitudes toward risk. One is the so-called "minimax" criterion, which is to choose the alternative for which the worst possible outcome is least unfavourable. In our example, spraying could result in a loss of \$119,224 while the maximum loss without spraying is only \$84,244, so the latter is preferred by this rule. This rule is suitable only for investors who are highly averse to risk; by focusing only on minimum possible returns and ignoring probabilities it biases decisions against attractive projects if they have even a small probability of poor returns.

A variety of other rules of thumb can be used to assist decision-making in the face of uncertainty. None can be said to be the correct rule in all cases because the best choice is that which best reflects the decision-maker's attitude toward risk. However, it is important for decision-makers to select a specific decision rule so that all the relevant information can be brought to bear appropriately and decisions can be made consistently. Because uncertainty is inherent in forestry investment decisions it is particularly important to account for it explicitly and consistently in evaluation procedures.

SOCIAL CONSIDERATIONS

Much of the methodology for evaluating investment opportunities has been developed to assist private investors whose concerns are relatively narrowly focused on potential returns and associated risk. Public investments often raise additional complications.

First, public investors, like private investors, must recognize uncertainty and the risk attached to their decisions. But because governments typically make larger and more diverse investments they can spread the risk and thereby reduce it, by the process noted earlier.

Second, public investment is usually more broadly concerned with efficient use of resources generally, in contrast to private investors who may be interested in maximizing returns to their land, capital, or other particular assets. The appropriate criterion for establishing priorities for public investments is thus more likely to be the benefit-cost ratio.

Third, the benefits to be taken account of must correspond to the decision-maker's objectives, and these may be more diverse for public investments. Effects on employment, impacts on the long-term stability of regional resource supplies, consequences for other resource users, and changes in taxes and transfer payments may not bear on private decisions but may figure importantly in public decision-making.

Fourth, public investment projects often raise more complicated questions about whose benefits and costs are to be considered. Activities in one forest can inflict benefits and costs on owners of adjacent land, or on people whose interest in the forest is indirect, such as those who depend on the watershed downstream. These are externalities which are often ignored by private decision-makers but must be included in any analysis that takes the viewpoint of society as a whole.

This is the *referent group* problem. It draws attention to the importance of clearly defining the scope of the society whose costs and benefits are to be considered. It is especially important where governments of differing levels are concerned with the same project. For example, the benefits of a forestry project may accrue to local residents while its costs are borne largely by taxpayers outside the area. Then an evaluation of the benefits and costs to the local community might reveal substantial net benefits while an evaluation from the national viewpoint would indicate costs exceeding benefits. The appropriate referent group must therefore be defined for each governmental decision-maker.

The principles outlined in this chapter for analysing the relationships between costs and benefits over time are fundamental tools of forest economics, and are employed throughout the remainder of this book. In the following chapter they are applied to the special problem of identifying the most advantageous age to grow forest crops.

NOTES

1 Equation 5 can be obtained from this equation by multiplying all terms by $\frac{1}{1+i}$ and simplifying, i.e.:

$$V_o = \frac{a\left[\frac{1}{1+i}\right]}{1 - \frac{1}{1+i}}$$

$$= \frac{a}{i}$$

which is Equation 5.

2 Equation 6 is obtained by multiplying all terms in the equation by $\frac{1}{1+i}$ and simplifying, i.e.:

$$V_o = \frac{a\left[\frac{1}{1+i} - \frac{1}{(1+i)^{n+1}}\right]}{1 - \frac{1}{1+i}}$$

$$= \frac{a\left[1 - \frac{1}{(1+i)^n}\right]}{i}$$

$$= \frac{a[(1+i)^n - 1]}{i(1+i)^n}$$

3 The simplification involves multiplying all terms by $1 \div (1+i)^n$ to obtain

$$V_o = \frac{V_n\left(\frac{1}{(1+i)^n}\right)}{1 - \frac{1}{(1+i)^n}}$$

APPENDIX A

Table of compound factors showing some values of $(1+i)^n$

Years	Interest rate (i) in per cent										
(n)	.05	1	1.5	2	3	4	5	6	8	10	15
1	1.005	1.010	1.015	1.020	1.030	1.040	1.050	1.060	1.080	1.100	1.150
2	1.010	1.020	1.030	1.040	1.061	1.082	1.102	1.124	1.166	1.210	1.322
3	1.015	1.030	1.046	1.061	1.083	1.125	1.158	1.191	1.280	1.231	1.521
4	1.020	1.041	1.061	1.082	1.126	1.170	1.216	1.262	1.360	1.464	1.749
5	1.025	1.051	1.077	1.104	4.159	1.217	1.276	1.838	1.409	1.611	2.011
6	1.030	1.082	1.093	1.126	1.194	1.265	1.340	1.419	1.587	1.772	2.313
7	1.036	1.072	1.110	1.149	1.230	1.316	1.407	1.504	1.714	1.949	2.660
8	1.041	1.083	1.126	1.172	1.267	1.369	1.477	1.594	1.831	2.144	3.059
9	1.046	1.094	1.143	1.195	1.305	1.423	1.551	1.689	1.999	2.358	3.518
10	1.051	1.105	1.161	1.219	1.344	1.480	1.629	1.791	2.159	2.594	4.046
11	1.056	1.116	1.178	1.243	1.384	1.539	1.710	1.898	2.832	2.853	4.652
12	1.062	1.127	1.196	1.268	1.426	1.601	1.796	2.012	2.518	3.188	5.350
13	1.067	1.138	1.214	1.294	1.469	1.665	1.886	2.133	2.720	3.452	6.153
14	1.072	1.149	1.232	1.319	1.513	1.732	1.980	2.261	2.937	3.798	7.076
15	1.078	1.161	1.250	1.346	1.558	1.801	2.079	2.397	3.172	4.177	8.137
16	1.083	1.173	1.269	1.373	1.805	1.873	2.183	2.540	3.426	4.595	9.358
17	1.088	1.184	1.288	1.400	1.853	1.948	2.292	2.693	3.700	5.054	10.76
18	1.094	1.196	1.307	1.428	1.702	2.026	2.407	2.854	3.996	5.560	12.38
19	1.099	1.208	1.327	1.457	1.754	2.107	2.527	3.026	4.316	6.166	14.23
20	1.105	1.220	1.347	1.486	1.806	2.191	2.653	3.207	4.681	6.728	16.37
21	1.110	1.232	1.367	1.516	1.860	2.279	2.786	3.400	5.034	7.400	18.82
22	1.116	1.245	1.388	1.546	1.916	2.370	2.825	3.604	5.437	8.140	21.64
23	1.122	1.257	1.408	1.577	1.974	2.465	3.072	3.820	5.871	8.954	24.89
24	1.127	1.270	1.429	1.608	2.033	2.553	3.225	4.049	6.341	9.850	28.63
25	1.133	1.282	1.451	1.641	2.094	2.666	3.386	4.292	6.848	10.83	32.92
30	1.161	1.348	1.563	1.811	2.427	3.243	4.322	5.744	10.06	17.45	66.21
35	1.191	1.417	1.684	2.000	2.814	3.946	5.516	7.686	14.79	28.10	133.2
40	1.221	1.489	1.814	2.208	3.262	4.801	7.040	10.29	21.72	45.26	287.9
45	1.252	1.565	1.954	2.438	3.782	5.841	8.985	13.76	31.92	72.89	538.8
50	1.283	1.645	2.105	2.692	4.384	7.107	11.47	18.42	46.90	117.4	-
55	1.316	1.729	2.288	2.972	5.082	8.646	14.64	24.65	68.91	189.1	-
58							***16.94***[a]				
60	1.349	1.817	2.443	3.281	5.892	10.82	18.68	32.99	101.3	304.5	-
65	1.383	1.909	2.632	3.623	6.830	12.80	23.84	44.14	148.8	400.4	-
70	1.418	2.007	2.835	4.000	7.918	15.57	30.43	59.06	218.6	789.7	-
75	1.454	2.109	3.065	4.416	9.179	18.95	38.83	79.06	321.2	-	-
80	1.490	2.217	3.291	4.875	10.64	23.05	49.56	105.8	472.0	-	-
85	1.528	2.330	3.545	5.883	12.34	28.04	63.25	141.6	883.5	-	-
90	1.587	2.449	3.819	5.943	14.80	34.12	80.73	189.5	-	-	-
95	1.606	2.574	4.114	6.562	16.58	41.51	103.0	253.5	-	-	-
100	1.647	2.705	4.432	7.245	19.22	50.50	131.5	339.3	-	-	-

[a]Value for 58 years at 5% needed to solve problems in Chapter 7

APPENDIX B
Compounding and discounting formulae commonly used in forestry

Compounding and discounting single values:

- the amount to which a value will grow over a period

$$V_n = V_o (1+i)^n$$

- the present value of a future amount

$$V_o = \frac{V_n}{(1+i)^n}$$

Discounting an annual series of values:

- the present value of a perpetual annual series beginning in one year

$$V_o = \frac{a}{i}$$

- the present value of a terminating annual series beginning in one year

$$V_o = \frac{a[(1+i)^n - 1]}{i(1+i)^n}$$

Discounting a periodic series of values:

- the present value of a perpetual periodic series beginning in one period

$$V_o = \frac{V_t}{(1+i)^t - 1}$$

- the present value of a terminating periodic series beginning in one period

$$V_o = \frac{V_t[(1+i)^n - 1]}{(1+i)^n[(1+i)^t - 1]}$$

Compounding a series of values:

- the final value of an annual series beginning in one year

$$V_o = \frac{a[(1+i)^n-1]}{i}$$

- the final value of a periodic series beginning in one period

$$V_o = \frac{V_t[(1+i)^n-1]}{(1+i)^t-1.}$$

Definition of terms:

i = annual rate of interest expressed as a decimal fraction
n = number of years of compounding or discounting
a = value recurring annually
t = number of years between periodic recurrences of V_t
V_t = value recurring periodically at intervals of t
V_o = present or inital value (at beginning of year 1)
V_n = future or final value (at end of year n)
annual = each year
periodic = at intervals of two or more years (t).

REVIEW QUESTIONS

1 Why is $100 received today worth more than $100 to be received five years hence? Given an interest rate of 10 per cent, how do these two payments compare in value? By how much would the future receipt have to be increased to make it equivalent in value to the $100 today?
2 What is the present value of a forest park that generates recreational benefits of $250,000 per year, given an interest rate of 7 per cent?
3 A developer has a standing offer to purchase a parcel of forest land for $1200 per hectare. The forest supports a forty-year-old crop that would yield $3000 per hectare if harvested in ten years and a similar amount, net of all costs, every fifty years thereafter. If the forest owner's interest rate is 8 per cent, is it advantageous to him to sell the land to the developer now, after harvesting the present crop in ten years, or never?
4 Recalculate the optimal level of fertilization in Review Question

number 4 in Chapter 2, assuming that the timber values would be realized five years after the fertilizer is applied, given an interest rate of 4 per cent.

5 If you had a limited budget, and you wanted to allocate it among a variety of potential silvicultural projects to maximize the economic benefits, what criterion would you use to establish priorities among projects?

6 What is the real rate of interest if the market or nominal rate is 12 per cent and the inflation rate is 4 per cent?

FURTHER READING

Baumol, W.J. 1968. On the social rate of discount. *American Economic Review* 58(40):788-802

Davis, L.S., and K.N. Johnson. 1987. *Forest Management*. 3rd ed. New York: McGraw-Hill. Part 2

Gunter, John E., and Harry L. Haney, Jr. 1984. *Essentials of Forestry Investment Analysis*. Corvallis, OR: OSU Book Stores

Harou, Patrice A. 1982. Evaluation of forestry programs: the with-without analysis. *Canadian Journal of Forest Research* 14(4):506–11

Herfindahl, Orris C., and Allen V. Kneese. 1974. *Economic Theory of Natural Resources.* Columbus, OH: Charles E. Merrill. Chapter 5

Mishan, E.J. 1982. *Cost-Benefit Analysis: An Informal Introduction.* 3rd ed. London: George Allen & Unwin. Part 5

Pearce, D.W. 1983. *Cost-Benefit Analysis.* 2nd ed. London: Macmillan. Chapters 5–8

Row, Clark, H. Fred Kaiser, and John Sessions. 1981. Discount rate for long-term forest service investments. *Journal of Forestry* 79(6):367–9

Sugden, Robert, and Alan Williams. 1978. *The Principles of Practical Cost-Benefit Analysis.* Oxford: Oxford University Press. Chapters 4, 5, and 15

Thompson, Emmet F., and Richard W. Haynes. 1971. A linear programming-probabilistic approach to decision making under uncertainty. *Forest Science* 17(2):224–9

CHAPTER SEVEN

The Optimum Forest Rotation

Out in the Forest . . .

In all investment planning, Peavey Forest Products Limited aims at a rate of return of at least 12 per cent, because it can earn that much on investments outside the company. The company's biggest capital asset is its inventory of standing timber, and one of Ian Olson's main responsibilities as the company forester is to manage this inventory so that it will yield an adequate return.

Central to this task is his choice of the age at which stands of growing timber should be harvested. The choice determines how long each stand must continue to earn interest, and also how big the total inventory must be to sustain the company's annual harvest. Over the last two decades, as a result of improvements in silviculture, new utilization technology, and other changes, Olson has reduced the average crop rotation age from sixty-five to fifty years, improving considerably the economic performance of the forest enterprise.

In planning the rotation age for each stand, Olson considers its annual growth in volume and value at various ages. Against this gain he must weigh the annual cost of carrying the forest crop, which is the interest on the value of the timber, and the annual cost of tying up the land. To maximize returns he must harvest the crop at the age when the gain from postponing the harvest for another year no longer exceeds the cost. His calculations of this optimum rotation period show that it is economically advantageous to harvest stands on his most productive sites at a younger age than those on poorer sites.

One of the most critical economic questions in forestry is the age at which trees should be harvested, or the *crop rotation* period. The choice governs how long the capital tied up in the crop must be carried before it is liquidated, and it also governs the size of the forest inventory that must be carried to maintain a given level of production. It is a problem that calls for analysis of biological as well as economic relationships over time and, like the aging of wine, it has intrigued economists for more than a century as a classical problem in investment analysis.

Foresters have developed a variety of criteria for selecting the age to harvest forest stands, some of which take no account of the economic variables involved. Examples are the age at which the trees reach a size best suited for making certain products, the age at which the volume in the stand is maximized, and the age at which the rate of growth in volume is maximized. These technical criteria are likely to prescribe widely divergent rotation ages, with major implications for the economic costs and benefits generated. Here we are concerned with finding the age that will yield maximum economic returns, taking account of how the volume and technical characteristics of the forest change with age and are reflected in its economic value.

Determining the most economically advantageous crop rotation period can be regarded as an investment problem of the kind discussed in the preceding chapter. To maximize the net benefits from growing a crop of timber we must examine how the recoverable values and the costs of producing them differ at different ages. Since the costs and benefits associated with a forest crop accrue at different times, both must be discounted to their present equivalent values so they can be compared consistently. Then the age which shows the greatest difference between the present value of benefits and the present value of costs can be identified. This chapter demonstrates how this economically optimum rotation age can be determined, and how it varies according to the biological growth and economic characteristics of the forest.

SOME INITIAL SIMPLIFICATIONS

To simplify the problem, we assume to begin with that the only benefit of concern to the forest manager is commercial timber. Recreational benefits, wildlife, livestock forage, and the host of other non-timber values that are important considerations in particular circumstances will initially be set aside. Later in the chapter we will note how some of these other forest values may call for modification of the silvicultural regime and the rotation period.

It is also convenient to assume that the appropriate management regime involves clearcutting the entire crop when it reaches harvesting age. This means we are concerned with determining the rotation period of even-aged forests, where the rotation period is the number of years between complete harvests of all the trees on the site.

We will begin with the simplest possible case: where the crop is established without cost and without delay; where only one crop is to be considered, after which the land will be valueless; where no taxes or management costs will be incurred as the crop grows; and where all costs and prices remain constant during the growing period. Later, these assumptions are relaxed in order to examine more realistic circumstances.

STUMPAGE VALUE AND STAND AGE

The value of the timber in a forest stand is referred to as its *stumpage value*. It is the maximum price that competitive buyers would be prepared to pay for the timber standing in the forest. Accordingly, the stumpage value, S, is equal to the revenue, R, that an efficient producer could expect from harvesting the timber and selling it in the best available market, minus his expected costs, C, in harvesting the timber and delivering it to the market.

$$S = R - C \tag{1}$$

The costs must include both capital and operating costs and a normal profit to the producer. Stumpage value thus embodies all the revenues and costs that guide the forest manager in his harvest planning.

A forest stand, growing on a hectare of land, increases in stumpage value as its age increases, following the general pattern indicated by the curve S(A) in the upper quadrant of Figure 19, where A refers to the age of the stand. Stumpage value increases with stand age for at least three reasons. First, the volume of merchantable timber on the hectare increases as the trees grow. The growth in volume of the stand is illustrated by the dashed curve Q(A) in the upper quadrant of Figure 19. It follows a sigmoid curve, its slope increasing up to an inflection point then decreasing, a growth pattern frequently observed in biology. In the case of a forest stand, the volume continues to increase as long as the (diminishing) annual increment of growth exceeds the (increasing) losses due to insects, disease, and natural mortality in the stand.

Another reason why the stumpage value of a stand increases as it grows older is that the trees become bigger, and more valuable products can be manufactured from larger timber. For example, large-dimension lumber and high-quality veneer can only be milled from large logs. Moreover, a larger proportion of the wood in large logs has a clear grain. Such quality differences usually cause the value of a stand of timber *per cubic metre* to rise as the trees grow bigger with age.

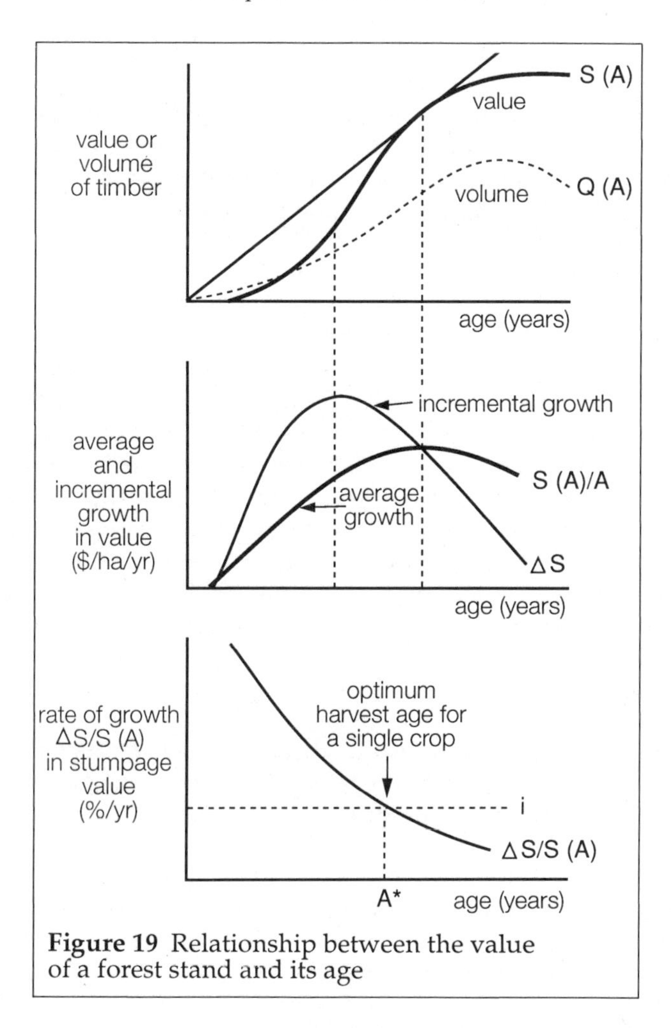

Figure 19 Relationship between the value of a forest stand and its age

A third reason is that larger timber can usually be harvested at lower cost *per cubic metre*, reflecting economies of log size. Large pieces require less handling per cubic metre because more wood can be yarded per turn and more volume loaded onto logging trucks than if pieces are small. Consequently, logging costs per cubic metre are lower the older and larger are the trees being harvested. (There are exceptions to this rule; for example, the uniformity in size of the logs is often as important an influence on handling costs as their average size.)

The effect of these influences on the value of a forest is that the growth in stumpage value usually follows a pattern similar to the growth in volume but at a faster rate and over a longer period, as shown in the upper quadrant of Figure 19.

With information about the stumpage value of a stand at various ages, reflected in the curve S(A), it is a simple matter to calculate the *average rate of growth* in the stumpage value of the stand to any chosen age. It is simply the value of the stand at a particular age divided by the number of years to that age, that is, S(A)/A; geometrically its value is the slope of a ray drawn from the origin of the upper quadrant in Figure 19 to the point on the total stumpage value curve corresponding to the relevant age. Its value, over the corresponding range of stand age, is shown in the middle quadrant of Figure 19.

The *incremental rate of growth* (ΔS) is the increase in stumpage value from one year to the next, that is, $\Delta S = S(A+1) - S(A)$. It represents the incremental gain in value of the stand if the harvest is delayed an additional year, a gain which varies with the age of the stand.

Geometrically, this annual incremental growth is represented by the slope of the stumpage value curve, S(A), in the upper quadrant in Figure 19. It therefore increases up to the inflexion point on the stumpage value curve, and then falls, as shown in the middle quadrant of Figure 19.

The relationship between the curves of average and incremental value growth corresponds to the relation between average and marginal cost curves in the economic theory of the firm. As long as the increment in value growth from one year to the next is greater than the average growth in value to that age, the average curve must continue to rise. At its maximum the average and incremental value growth are equal, and it declines over the range where the incremental growth is less than the average, as shown in Figure 19.

OPTIMUM ECONOMIC ROTATION

The incremental growth in value of the stand, expressed as a percentage of the current stumpage value, ΔS/S(A), follows a pattern illustrated in the lower quadrant of Figure 19. It declines as the stand ages, because the denominator increases and the increment in value growth declines over a wide range.

A forest owner wanting to select the most advantageous harvesting age must consider the marginal benefit and marginal cost of carrying the crop from one year to the next. More specifically, he must, in any year, weigh the return on his capital by growing the crop for another year, ΔS/S(A), against the cost of doing so. Ignoring for the moment the cost of the land, the owner's cost in carrying the crop is the interest he could earn on the capital tied up in the crop if he liquidated it and invested the proceeds at the going rate of interest, i. So, to maximize his gains, he will carry the crop only so long as the rate of return on growing the stand exceeds the interest rate, which is to say he will carry the crop so long as the marginal benefit from doing so exceeds the marginal cost, and no longer. The age at which the return falls to the interest rate, $\Delta S/S(A) = i$, is therefore our first approximation to the optimum rotation age, denoted A* in the lower quadrant of Figure 19.

With these simplest possible assumptions, the rule for the optimum rotation age is to equate the marginal benefit of carrying the forest capital for another year, the percentage increase in stumpage value, with the opportunity cost of the capital.

$$\Delta S = i(S(A)) \qquad (2)$$

This implies that the crop should be carried until the incremental rate of return from another year's growth falls to the opportunity cost of capital. The optimum economic rotation is therefore *longer the higher and more prolonged the rate of stand growth and the lower the rate of interest.*

This solution is consistent with the maximization of the present worth of the harvest, following the investment analysis in the previous chapter. Following the formulation in Chapter 6 (Equation 2), the present net value (V_o) of the harvest, discounted to the beginning of the crop rotation, is

$$V_o = \frac{S(A)}{(1+i)^A}. \qquad (3)$$

The optimum economic rotation is the age at which this value is greatest. With information about the rate of growth in stumpage value of the stand and the opportunity cost of capital, this optimum age can be calculated as the age at which the annual increase in present value of the stand falls to zero,

$$\Delta V_o = 0$$

or

$$\frac{S(A+1)}{(1+i)^{A+1}} - \frac{S(A)}{(1+i)^A} = 0.$$

This equation is equivalent to

$$S(A+1) = (1+i)(S(A))$$

or

$$\Delta S = i(S(A))$$

which is the same as the rule in Equation 2.

OPTIMUM ROTATION WITH SUCCESSIVE CROPS

The above solution takes account of only one cost, namely that of tying up capital. But in forestry there are always at least two factors of production involved, capital and land, and hence two categories of costs to be considered. The cost of the land therefore must be incorporated into our analysis of the optimum economic rotation.

Let us assume for the moment that the land is suitable only for growing timber, or finds its most productive use in this activity, and that each successive crop will involve identical values and costs. The net present value, V_o, of an infinite series of future harvests, S(A), expected at regular intervals, A, can be expressed as the geometric series

$$V_o = \frac{S(A)}{(1+i)^A} + \frac{S(A)}{(1+i)^{2A}} + \frac{S(A)}{(1+i)^{3A}} + \ldots + \frac{S(A)}{(1+i)\infty}$$

where each successive term on the right-hand side represents the present net worth of another crop after an additional rotation period of A years. Adapting the formula derived in the preceding chapter, this expression can be simplified to

$$V_o = \frac{S(A)[1/(1+i)^A]}{1-1/(1+i)^A}$$

or its equivalent

$$V_o = \frac{S(A)}{(1+i)^A-1.}$$

This present worth of an infinite series of future harvests, net of all costs of producing them, is sometimes referred to as the "site expectation value," "soil rent," or "bare land value"; here we will use the term *site value*, V_s. If there are no costs involved in producing the crop, the site value can be expressed as

$$V_s = \frac{S(A)}{(1+i)^A-1} \quad (4)$$

The site value is the value of the land for purposes of continuous forestry, evaluated when it is in a bare state at the beginning of a rotation.[1]

The optimum economic rotation for a continuing succession of crops is that which will generate the highest site value. This is the age at which the present net worth of the forestry enterprise cannot be increased by extending the rotation age by another year, that is,

$$\Delta V_S = O,$$

which means that

$$\frac{S\,(A)}{(1+i)^A-1} = \frac{S\,(A+1)}{(1+i)^{A+1}-1}\,,$$

which can be simplified to[2]

$$\frac{\Delta S}{S\,(A)} = \frac{i}{1-(1+i)^{-A}}. \tag{5}$$

At the optimum forest rotation, A**, this equality will be satisfied. Again, it implies balancing the marginal benefit, expressed as the percentage increase in stumpage value from carrying the crop another year (the term on the left-hand side of Equation 5) with the marginal cost, including the annual cost of holding the land (the term on the right-hand side). At any rotation age less than A**, it is advantageous to postpone harvesting because the incremental gain in value exceeds the incremental cost, and at any greater age the incremental cost exceeds the gain in value, as shown in Figure 20. This expression for the optimum economic rotation for continuous forestry is the *Faustmann formula*, having been derived by the German capital theorist Martin Faustmann in 1849.

Figure 20 illustrates the relationships between the terms on each side of Equation 5, and the optimum rotation at age A** where they are equal.

As shown in Figure 20, the optimum rotation age, A**, when successive crops are considered, is shorter than the optimum rotation, A*, for a single crop. Algebraically, this is because the incremental cost under successive crops, $i \div [1-(1+i)^{-A}]$ is greater than the incremental cost, i, for a single crop (because the denominator in the former is less than 1). Graphically, this means that the curve of incremental cost under successive crops is higher, and therefore intersects the curve of incremental growth in value at an earlier rotation age. As Figure 20 shows, the incremental cost $i \div [1-(1+i)^{-A}]$ exceeds the interest rate, i, and becomes asymptotic to it at high rotation ages.

Logically, the shorter rotation under successive crops can be explained by the addition of the cost of the second factor of production in forestry, land, which raises the incremental cost of postponing the harvest and causes it to intersect with the incremental growth in value at an earlier age. The opportunity cost of the land,

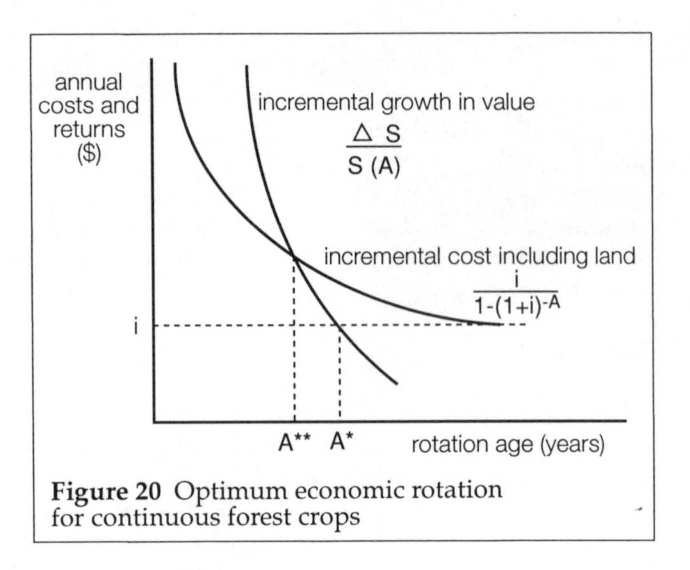

Figure 20 Optimum economic rotation for continuous forest crops

which measures the value it could generate if the crop were harvested and a succession of new crops initiated, adds urgency to the harvest. Each year the present crop is carried the value of future crops, represented by the site value, is postponed another year, so it is advantageous to harvest at an earlier age.

However, the extent to which the optimum rotation is shortened by taking account of subsequent crop values in this way may be small because the force of discounting reduces the value of later harvests, especially over long rotation periods and at high discount rates. For example, a single crop worth $10,000 in sixty years has a present value, discounted at 5 per cent using Equation 3, of $535. The present value of an infinite succession of crops every sixty years, each valued at $10,000, discounted at the same rate using Equation 4, is only $30 greater. This is the present value of all the harvests following the first one. Consequently, the economic rotation is only slightly shorter when more than one crop is considered, but the effect is greater the lower the discount rate and the shorter the rotation age.

The incremental cost of carrying the forest varies with the interest rate, whether only one or a succession of crops is considered. Thus, the higher the interest rate the shorter the optimum rotation.

AN ILLUSTRATION

The way these economic variables converge to define the optimum economic rotation can be illustrated with reference to Table 2, which shows how a particular forest type grows in volume and value over time. The stumpage values of the stand, S(A), in the fourth column enable us to calculate the site value at different ages using Equation 4. Using a discount rate of 5 per cent, the maximum site value of $827 per hectare occurs at a rotation age of fifty-eight years (based on straight-line interpolation between the values for 55 and 60). The annual opportunity cost of the land, or the land rent, a, is the annual equivalent of this site value

$$\begin{aligned} a &= iV_s \\ &= .05\,(827) \\ &= \$41.4, \end{aligned}$$

which is the highest value shown in the eighth column of Table 2. This maximum site value and its equivalent annual land value define the optimum economic rotation.

A forest owner must consider this annual land cost as well as the annual interest on the stumpage value of the crop, iS(A) in the sixth column of Table 2, in deciding his rotation age. The opportunity cost of the land and capital required to carry the crop from one year to the next is what they could earn if he liquidated the crop. Harvesting would release the capital embodied in the timber, S(A), which has an annual value of iS(A). It would also release the land, which would then have a value equal to the site value, V_S, and an annual equivalent value, a.

Against these two incremental costs of carrying the crop, he must weigh the incremental gain in stumpage value ΔS(A). Thus, the optimum rotation age, A**, is

$$\Delta S = iS(A) + a \tag{6}$$

This is the same as the Faustmann formula in Equation 5.[3] It implies that the owner will continue to grow the crop as long as the increment in value exceeds the increment in costs, but no longer.

Using the data in Table 2 these relationships are illustrated in Figure 21. This shows the incremental growth, ΔS, exceeding the incremental costs, iS(A) + a, up to fifty-eight years, indicating the optimum economic rotation.

Once again, the optimum economic rotation is defined, as in Equation 6, as the age at which the annual incremental growth in the stumpage value of the crop is just equal to the incremental costs in carrying the crop. The incremental costs include not only the interest on the capital embodied in the crop but also the interest on the

TABLE 2: Value and costs of growing a forest to various harvesting ages

Age	Volume per hectare[a]	Value per cubic metre[b]	Stumpage value per hectare	Annual incremental change in stumpage value	Annual interest on stumpage value[c]	Site value[d]	Land rent[e]	Annual incremental cost[f]
A	Q(A)		S(A)	ΔS(A)	iS(A)	V_s	a	iS(A) + a*
Years	m³	$	$	$	$	$	$	$
10	2	0	0		0	0	0	41
				0				
15	14	0	0		0	0	0	41
				0				
20	51	0	0		0	0	0	41
				0				
25	124	0	0		0	0	0	41
				93				
30	232	2	464		23	140	7.0	64
				200				
35	366	4	1,464		73	324	16.2	114
				323				
40	513	6	3,078		154	510	25.5	195
				444				
45	662	8	5,296		265	663	33.2	306
				545				
50	802	10	8,020		401	766	38.3	442
				626				
55	929	12	11,148		557	817	40.9	598
				680				
58	***995***	***12***	***13,187***	***690***	***659***	***827***	***41.4****	***700***
60	1,039	14	14,546		727	823	41.1	768
				713				
65	1,132	16	18,112		906	793	39.6	947
				730				
70	1,209	18	21,762		1,088	739	37.0	1,129
				732				
75	1,271	20	25,420		1,271	672	33.6	1,312

NOTES

[a] Volume for Douglas-fir in coastal British Columba, site index 40 at reference age 50 years. Figures rounded to nearest cubic metre. Source: B.C. Ministry of Forests, Inventory Branch. *Variable Density Yield Tables and Equations for Coastal Douglas-fir in British Columbia.* Forest Inventory Report No. 2. Ministry of Forests. Victoria. July 1982.

[b] Values based on assumed stumpage values increasing by $2 per m³ every 5 years beginning at age 25

[c] Using an interest rate of 5%

[d] Where $V_s = \frac{S(A)}{(1+i)^A - 1}$

[e] Where $a = iV_s$

[f] I.e., the sum of the annual interest on stumpage value, iS, and the maximum annual opportunity cost of land, a* ($41.2, at age 58), rounded to the nearest dollar

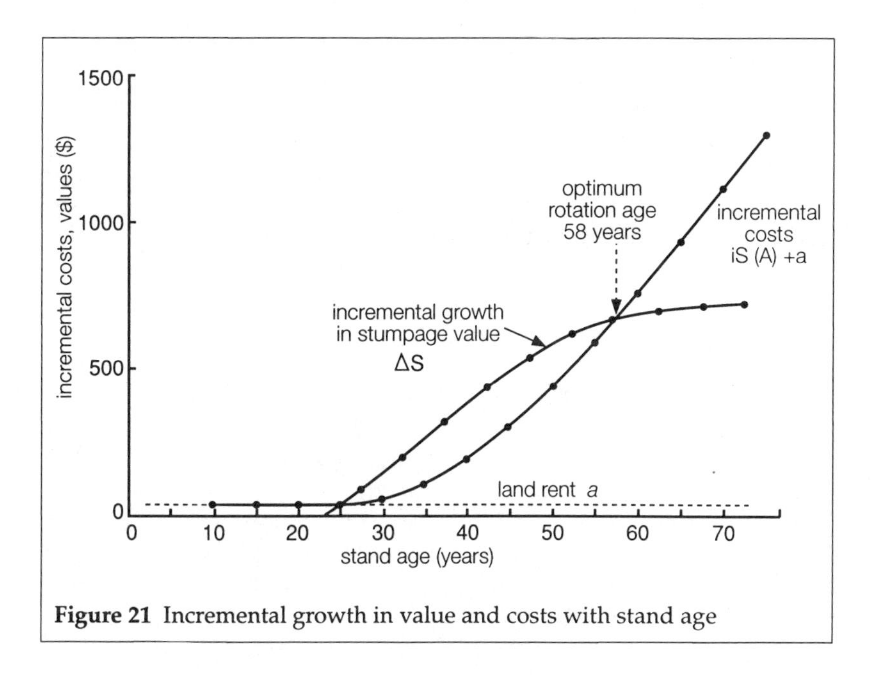

Figure 21 Incremental growth in value and costs with stand age

bare land value, or the annual rent, that the land is capable of generating in continuing forestry.

COMPARISONS WITH OTHER ROTATION CRITERIA

The Faustmann formula maximizes the economic rent to the fixed factor of production, land, and thereby maximizes the present net value of the future stream of forest crops. It follows that any other rotation age will yield lower net returns.

In the example presented in Table 2 and Figure 21, the optimum rotation age occurs before the incremental growth in stumpage value, ΔS, reaches its maximum, but in other cases it may be later. Moreover, it will usually differ from the age at which the average rate of growth in stumpage value, $S(A)/A$ (sometimes referred to as the "forest rent") is greatest. This can be readily seen in Figure 19. The significance of this lies in the historical popularity of "forest rent" as the guide to the optimum rotation. By maximizing the average annual growth in stumpage value, it is argued, the revenue from the forest is maximized. But "forest rent" (even though it is usually calculated net of cash costs of managing crops) ignores the opportunity costs of capital and land, and so maximizing "forest rent" does not maximize the net present value of forest crops.

The optimum economic rotation is also likely to differ substantially from the age that maximizes the average rate of growth in stand volume, Q(A), in the upper diagram of Figure 19, because the patterns of growth in value and volume differ, as noted earlier. This last difference is important because the age of maximum mean annual increment is a popular choice of rotation age in traditional forestry regulation. It usually indicates a longer rotation than the economic optimum. However, the differences in rotation age, and the extent to which potential returns are eroded by adopting the maximum mean annual increment criterion, vary widely.

OTHER IMPACTS ON THE OPTIMUM ROTATION

The optimum forest rotation is influenced by the productivity of the land, the value of the timber produced, the costs of harvesting it, taxes and other costs of management, the interest rate, and other conditions. These vary widely in different circumstances. The remainder of this chapter considers how some of these factors affect the optimum forest rotation.

Rate of Interest

A higher rate of discount will increase the incremental cost of carrying the forest crop and thereby shorten the optimum rotation, as can readily be seen in Equation 6 and Figure 21. This is partially offset by the lower land rent that results from heavier discounting, but the net effect is always an increase in cost, reducing the optimum rotation.[4] Thus *the higher the interest rate the shorter the optimum rotation*.

The interest rate is never as low as zero, for reasons explained in the preceding chapter, but it is instructive to consider such an extreme case. Then, in the absence of the force of discounting, the optimum rotation would simply be the age at which the average rate of growth in stumpage value, S(A)/A is maximized, referred to earlier as the age of maximum "forest rent." As the middle diagram of Figure 19 shows, this is the age at which the average and incremental rates of growth are equal. However, as long as the interest rate exceeds zero the optimum economic rotation must be shorter than the age at which the "forest rent" is maximized.

Reforestation Costs

Preceding examples assumed that no costs were incurred in establishing new crops following harvests. If reforestation costs must be incurred at the outset of each rotation, the rate of incremental growth in stumpage value, $\Delta S/S(A)$ in Figure 20, must be revised to $\Delta S/[S(A) - C]$, reducing the value of the crop by the cost, C, of replanting.[5] The effect will be to shift the incremental growth curve upward to the right, lengthening the optimum rotation, and *the higher the reforestation cost the longer the optimum rotation*.

Where planting is necessary, but no other costs are involved, growing successive crops will be advantageous if the value of the harvest, discounted to the beginning of the rotation, exceeds the planting cost. If this is so for one crop, it will be the case for successive crops, notwithstanding the impact of discounting.

Note that if the reforestation cost were equal to the present worth of the harvest (i.e., $C = S(A)/(1+i)^A$, so that the present net worth of growing a new crop were zero, it can be seen by substituting in Equation 5 that the optimum rotation would be the same as in the case of a single crop, derived from Equation 2. This is because the land would yield no net value in subsequent crops.

Land Productivity

Not only reforestation costs, but anything that raises the cost of growing the crop has the effect of lengthening the optimum forest rotation. The same is true of anything that lowers the value of the harvest or reduces the productivity of the land in terms of its capacity to generate value. All these reduce the value of the crop and hence reduce also the two incremental costs in Equation 6, the interest on the forest capital, $iS(A)$, and the land rent, a. They also reduce the incremental growth in value, ΔS, though by a smaller amount. The result can be seen in Figure 21; the incremental cost and benefits will intersect later, indicating a longer optimum rotation age, A^{**}. In other words, as long as other influences remain the same, *the more productive the forest site the shorter the optimum rotation*.

Annual Costs

Forestry usually involves some continuing costs of management, protection, administration, and taxes. If these costs are a constant annual amount (m) they can easily be provided for in the above formulae. As we saw in the preceding chapter, the present worth of such an infinite future series is simply m/i, which can be subtracted from the right-hand side of Equation 4, reducing the site expectation value by this amount.

Constant annual costs will not alter the optimum rotation, however, because the age that maximizes the site value will also maximize the site value minus a constant m/i.

This illustrates a general rule: *the optimum rotation is not affected by any cost that is independent of the way the forest is managed*. Any such cost will reduce the net returns from the forestry enterprise, of course, but as long as the burden cannot be shifted or lessened by changing the management regime it will not affect the rotation age.

Property Taxes

Taxes on forest properties usually consist of a percentage rate applied against the tax base, which is typically the value of the land, the value of the timber, or both of these combined.

Forest taxes and other charges are dealt with in detail in Chapter 10; here we briefly note their impact on the optimum forest rotation.

Tax on land value. The simplest property tax is one based on the bare land, usually consisting of a percentage tax rate applied annually to the site value. This produces a fixed annual charge and has the same effect as other annual costs noted above. Although it will lower the land's value to the owner, *a tax on the bare land value will not alter the most advantageous harvest age*. It will simply divert some of the value generated by the land from the owner to the government.

Tax on timber value. A tax rate applied annually to the value of the standing timber will result in an annual charge equal to a constant fraction of the stumpage value. The forest owner thereby will incur an annual cost, increasing each year with the stumpage value of the stand, and cumulating at compound interest until the stand is harvested. The return from growing the forest is the stumpage value when it is harvested minus the accumulated tax payments.

Such an annual *ad valorem* (or percentage of the value) tax on growing timber has the same effect on the owner's choice of the rotation age as an increase of that percentage in the rate of interest in the *Faustmann formula*. As already seen, increasing the discount rate shortens the optimum rotation. Correspondingly, the longer the crop is grown, the greater the accumulated taxes that must be charged against the harvested timber. This means that *an annual tax on the value of the standing timber will provide an incentive to shorten the rotation*.

Tax on harvest value. A yield tax, in the form of a percentage rate applied against the stumpage value, is sometimes levied on timber when it is harvested. A tax of this kind has the effect of reducing the value of the harvest to the owner in proportion to the tax rate.

Like higher reforestation costs that must be incurred when the forest is harvested, a yield tax can be postponed, and thereby reduced in present value, by postponing the harvest. The effect of this, and the reduced stumpage return to the owner, means that *a yield tax on the harvest lengthens the preferred rotation*.

Other Forest Values

As noted above, other forest products and services are affected by the way timber is managed and harvested. In some cases these other benefits are sufficiently valuable, and are so adversely affected by timber production, that any timber production would lower the forest's aggregate net return. Then economic efficiency, and maximization of land rent, precludes timber production. However, where returns can be increased by combining timber production with other uses of the forest, multiple use management plans must take account of the impact on other forest values of timber production, including the choice of harvesting age.

Some non-timber values are higher when the forest cover is young, or discontinuous, or has a high proportion of recently harvested areas. Examples are certain kinds of wildlife or grazing values. Other values are higher when forest stands are older, such as certain recreational and aesthetic benefits. Each forest product or service is affected by the age of the forest in a different way. But if they are affected at all they will affect, in turn, the rotation age that will maximize the aggregate net value of the forest.

The tables and figures earlier in this chapter illustrate how the value of timber changes with the age of the forest stand. Correspondingly, other values change with stand age, but they may follow a quite different pattern, increasing or decreasing with the age of the stand. To take account of non-timber benefits in determining the forest rotation, schedules of their values at various stand ages (corresponding to stumpage values at various ages) can be used to calculate their annual incremental change in value over the range of stand ages. This incremental value can then be added to the incremental growth in stumpage value, ΔS, to calculate the optimum economic rotation for the combined production of timber and non-timber values.

The effect can be readily seen with reference to Figure 21. If the non-timber value under consideration is greater in older forests, its annual incremental growth in value will be positive, so adding it to the incremental growth in stumpage value, ΔS, will raise this curve and lengthen the optimum rotation. If the non-timber value is greater in younger forests, its incremental growth with stand age will be negative, shortening the optimum rotation.

Trends in Costs and Prices

The discussion so far has implied that the stumpage values and costs that enter into the calculation of the optimum rotation are constant over time. In reality, however, they are likely to change. While it is often difficult to predict the direction and magnitude of future changes in these economic variables, we must incorporate our expectations, or best guesses, into our calculations.

Values and costs fluctuate constantly with changing economic conditions, but for purposes of planning forest crops over long periods the concern is not with short-term swings but with long-term trends. The costs and values used in the Faustmann formula must be adjusted to reflect expected trends over the long periods for which the calculations are made.

If stumpage values are expected to rise over time at some percentage rate, for example, this can be allowed for by reducing the discount rate applied to stumpage values by that percentage.[6] Expected trends in costs can be accommodated in a similar way.

This simple procedure of adjusting the discount rate to account for trends in values and costs can be used only when the trends are expected to be steady. If they are expected to change in some irregular fashion, each term in the geometric series that leads to Equation 4 will differ, so they cannot be reduced to a simple expression. Instead, each term must be treated separately, making the calculation of the optimum rotation much more cumbersome.

A variety of other costs, values, and taxes associated with growing timber are examined in later chapters. Most have some effect on the optimum rotation; as we have seen, only charges or revenues that are independent of the current value of the crop will have no effect. In most cases their impacts can be assessed by modification of the Faustmann formula.

The following chapter explains the important implications of the choice of rotation age for the scheduling of harvests in regulated forests.

NOTES

1 The site value, V_s, of bare land that must be planted at a cost, c, at the beginning of each rotation becomes

$$V_s = \frac{S(A) - c(1+i)^A}{(1+i)^A - 1}.$$

If an annual management cost, m, must be accounted for as well, $\frac{m}{i}$ must be subtracted from this value.

A convenient general formulation for calculating the present value of a perpetual forest where the crop is already partly grown and a variety of revenues R_1, R_2 ... and costs C_1, C_2 ... are expected at different times during the rotation, and management costs, m, are expected annually, is

$$V_o = \frac{R_1(1+i)^{A-T_1} + R_2(1+i)^{A-T_2} + \ldots - C_1(1+i)^{A-T_1} - C_2(1+i)^{A-T_2} - \ldots}{(1+i)^A - 1} - \frac{m}{i}$$

where T is the number of years from the present to the time the periodic cost or revenue will occur. The numerator of the first term on the right-hand side sums the value of all revenues and costs at the end of the current rotation, while the denominator provides for the perpetual periodicity of this value. The second term accounts for the present value of the annual management cost. In the special case where the stand is age zero, T is also the age of the stand, and V_o is the site value, V_s corresponding to equation (4).

2 The expression

$$\frac{S(A)}{(1+i)^A - 1} = \frac{S(A+1)}{(1+i)^{A+1} - 1}$$

can be solved for S(A+1), and S(A) subtracted from both sides to give

$$\Delta S = S(A)\left\{\frac{(1+i)^{A+1} - 1}{(1+i)^A - 1}\right\} - 1$$

and the term in large brackets can be simplified to obtain Equation 5.

3 Thus, from Equation 6:

$$\Delta S = i\,(S\,(A) + V_s)$$
$$\Delta S = i\,(\,(S(A^{**}) + S(A^{**})/[(1+i)^{A^{**}} - 1]\,)$$

$= iS\,(A^{**})\,(1+i)A^{**}/[(1+i)A^{**} - 1]$
$= iS\,(A^{**})/[1 - (1+i)^{-A^{**}}]$

which is the same as Equation 5.

4 This conclusion is obvious from Equation 5, which shows that the incremental cost on the right-hand side of the equation must increase with any increase in i.

5 In this case the site value is

$$V_s = \frac{S(A) - C(1+i)^A}{(1+i)^A - 1}.$$

6 If stumpage values are expected to rise at a rate, r, the stumpage value of a harvest A years hence will be $(1+r)^A$ (S(A), and the present value of a future series of crops becomes

$$V_o = \frac{(1+r)^A\,S(A)}{(1+i)^A} + \frac{(1+r)^{2A}\,S(A)}{(1+i)^{2A}} + \frac{(1+r)^{3A}\,S(A)}{(1+i)^{3A}} \ldots$$

and the expression for the site value, Equation 4, becomes

$$V_o = \frac{S(A)}{\left\{\frac{1+i}{1+r}\right\}^A - 1.}$$

Thus the effective interest rate for calculating the optimum rotation is reduced to i–r ÷ i+r which, for practical purposes is equal to i+r. It is therefore sufficient to use a discount rate of i–r in the Faustmann formula to account for the upward trend in stumpage values.

REVIEW QUESTIONS

1 Why is the growth in stumpage value of a forest stand not precisely proportional to its growth in wood volume?
2 What are the incremental costs of carrying a forest from one year to the next? What is the incremental benefit? Explain why harvesting at the age when the incremental costs and benefit are equal will maximize the return to the forest owner.
3 The merchantable volume of timber on a hectare of forest increases with the age of the stand as in the following table. Assume all timber has a value of $5 per cubic metre, and the applicable interest rate is 6%.

Stand age (years)	Timber volume (m^3/ha)
15	0
20	50
25	100
30	240
35	400
40	530
45	640
50	730
55	760

Calculate the land rent and the optimum economic rotation under continuous forest production.
4 What is the effect on the optimum rotation age of (a) an increase in the interest rate, (b) reduced reforestation costs, and (c) higher annual costs of fire protection?
5 If the recreational value of a forest is higher the older the trees, what effect will this have on the optimum age for harvesting the timber?

FURTHER READING

Anderson, F.J. 1985. *Natural Resources in Canada: Economic Theory and Policy*. Toronto: Methuen. Chapter 7
Bentley, William R., and Dennis E. Teeguarden. 1965. Financial maturity: a theoretical review. *Forest Science* 11(1):76–87

Calish, S., R.D. Fight, and D.E. Teeguarden. 1978. How do nontimber values affect Douglas-fir rotation? *Journal of Forestry* 76(4):217–21

Clark, Colin W. 1976. *Mathematical Bioeconomics: The Optimal Management of Renewable Resources*. New York: John Wiley & Sons. Chapter 8

Faustmann, Martin. 1968. Calculation of the value which forest land and immature stands possess for forestry. In M. Gane (ed.) and W. Linnard (trans.). *Martin Faustmann and the Evolution of Discounted Cash Flow: Two Articles from the Original German of 1849*. Institute Paper No. 42, Commonwealth Forestry Institute, University of Oxford. Oxford: Commonwealth Forestry Institute

Hartman, R. 1976. The harvesting decision when a standing forest has value. *Economic Inquiry*, 14:52–68

Heaps, Terry. 1981. The qualitative theory of optimal rotations. *Canadian Journal of Economics* 14(4):686–99

Hyde, William F. 1980. *Timber Supply, Land Allocation, and Economic Efficiency*. Baltimore: Johns Hopkins University Press for Resources for the Future. Chapter 3

Pearse, P.H. 1967. The optimum forest rotation. *Forestry Chronicle* 43(2):178–95

Rideout, D. 1985. Managerial finance for silvicultural systems. *Canadian Journal of Forest Research* 15(1):163–66

Samuelson, Paul A. 1976. Economics of forestry in an evolving society. *Economic Inquiry* 14(4):466–92

Walter, G.R. 1980. Financial maturity of forests and sustainable yield concepts. *Economic Inquiry* 18(2):327–32

CHAPTER EIGHT

Regulating Harvests over Time

Out in the Forest . . .

Markets for forest products are cyclical, and the market for logs is particularly vulnerable to swings in demand and price. In the face of fluctuating log prices, Peavey Forest Products Limited president David Cameron continually adjusts the rate of production in order to realize maximum returns on the timber on the company's lands. He cannot simply cut at a constant rate, or cut every stand as it reaches the planned rotation age. Instead, he must expand production when prices rise and reduce output when prices fall.

The main decision Cameron must make is how much to adjust output when market conditions improve or worsen. Log production costs, per cubic metre, are lowest with production at the company's planned operating capacity; costs rise whenever production is significantly above or below that level. This constrains his scope for advantageous adjustments in production in the short run.

In the long term the company can change its production capacity by investing in more equipment and other capital facilities, and over an even longer period it can change the productive capacity of the forest itself through investments in silviculture. Thus the more time the company has to adjust to a market change, the more flexibility it has to respond.

However, the market in which the company sells its logs is supplied, as well, with logs produced on public lands, where sustained yield regulations require a more-or-less steady annual cut. This prevents the supply of public timber from responding fully to market changes, which in turn lowers the return on this timber and aggravates cyclical price fluctuations.

The sustained yield controls are intended to provide more stable regional employment and income. However, their stabilizing effect on Sundry Island has been limited over recent decades because increasing productivity has substantially reduced the labour required to produce the same quantity of timber. Moreover, the sustainable harvest itself has had to be changed as a result of more intensive utilization and silviculture. In any event the stability of the island's economy depends largely on trends and fluctuations in tourism, service industries, and private forest production like that of Peavey Forest Products Limited, all of which respond to market forces.

How a forest should be harvested over time is one of the most fundamental issues of forest management. Decisions about how fast harvesting is to take place, and how it is related to rates of growth, are the primary means of managing the structure and composition of a forest. Moreover, because harvesting is the activity that generates revenues and reduces the capital tied up in timber, its timing is critical to the economic performance of forest enterprises.

Even beyond the scale of the individual forest or firm, decisions about the harvest rate govern, in large part, the economic and social impacts of forestry. By determining regional timber supplies, the harvest rates chosen influence the size of the forest industry and its stability over time. And logging and manufacturing sectors, which are often the foundation of regional economies, must adapt their capacities accordingly. For these reasons decisions about the level of harvesting and its spread over time, examined in this chapter, are major preoccupations of both private forest owners and governmental forest agencies.

THE STAND AND THE FOREST

The preceding chapter dealt with the age at which a "stand" of timber should be harvested in order to maximize economic returns. This chapter is concerned with the "forest," which usually consists of many stands, each having different characteristics with respect to species composition, age, site productivity, and so on. While the previous chapter considered the time between harvests on a particular site, this chapter deals with the rate of harvesting in the whole forest.

A forest, delineated by natural or artificial boundaries, is the unit for most management decisions relating to the whole forest enterprise. Although managers must pay attention to the particular management problems and opportunities of each distinctive stand, major decisions about access development, investment in forest enhancement, protection, rates of harvesting, and the measurement of economic performance normally apply to the forest as a whole. This means that individual stands are managed with reference to the manager's objectives for the whole forest.

The benefits derived from a forest are not simply the sum of the benefits derived from all the component stands managed independently. This is because some management decisions affect many stands. For example, the location of roads and processing plants influence the cost of transportation and hence the value of timber in different areas. Another reason is that forestry activities such as fire protection, pest control, and silviculture need to be co-ordinated over many stands in order to be effective and to achieve economies of scale. The production of some goods and services other than timber also calls for co-ordinated management of many stands. Wildlife and scenic and recreational values, for example, require management on a larger scale than individual stands.

For all these reasons some forest management problems must be considered in the context of the whole forest. Among the most important of these is the question of how harvesting should be scheduled over time, the subject of this chapter. This question raises much broader issues than the narrow one of the age at which an isolated stand should be harvested to maximize returns to its particular site, considered in Chapter 7.

The relationship between this problem and the market supply of timber deserves emphasis. In Chapter 3 the market supply was depicted as the sum of the quantities of timber that all the producers were prepared to offer for sale in a particular market. Here we are concerned with how the manager of a particular forest determines his production. He is thus an individual producer who, usually with others, determines market supply.

MARKET SOLUTIONS AND LIMITATIONS

To a large extent, the problem of regulating the harvest rate is akin to the problem most firms face in managing an inventory. The stock of timber in the forest is costly to carry over time, its opportunity cost being the potential rate of return that could otherwise be earned on its capital value. Like other inventories the forest can be diminished through utilization and sale, and augmented by investment in more production.

In a perfectly competitive market economy like that described in Chapter 2, private forest owners could be expected to manage their forest inventories and timber supplies independently and efficiently in response to market costs and prices. The key to obtaining the greatest possible value from their forests has been explained in earlier chapters; they must be managed and harvested so that the net present value of the stream of goods and services produced over time is maximized. For any particular stand, this means that harvesting should occur when its growth in value, from one year to the next, no longer exceeds the incremental cost of carrying it for another year, as explained in Chapter 7 and Figure 21. If the forest owner had no concerns other than maximizing his economic return from the forest itself, and if he could alter his harvesting by any amount without affecting his costs and prices, he could simply apply this rule for the optimum rotation to every stand in his forest. He would thereby generate the maximum possible value from each stand and from the forest as a whole.

Our particular interest in this chapter is the yield of timber over time that would flow from the forest under this regime. The outcome obviously depends on the age distribution of the component stands. To illustrate the range of possibilities consider a forest made up of stands of uniform size and productivity, the number of stands equal to the number of years in the rotation period. If, at one extreme, there was a stand of each age class from zero to the rotation age, the rule for maximizing returns described in the preceding chapter would result in one stand being harvested each year forever. The yield would be constant over time. At the other extreme, if all the stands were the same age, the whole forest would be harvested in a single year in each rotation period.

In most cases the composition of forests lies somewhere between these extremes. Age classes are irregularly distributed, so the yield of timber harvested according to the efficiency rule would be irregular over time. Moreover, in practice the stands vary in size and productivity, and the number of them does not correspond in any way to the rotation age, all of which compound the irregularities resulting from an uneven age distribution.

The yield of timber over time would be affected by other destabilizing influences as well. In particular, fluctuating markets for timber would give producers incentives to accelerate harvesting when prices are high and to reduce them when prices are low. And over very long periods, changes could be expected in technology, costs, and prices which would significantly alter yields, silvicultural opportunities, optimum rotation periods, and other variables, and these changes would also affect yields over time.

The result of all these influences would be a changing level of production of timber from the forest over time, and the pattern of fluctuations would depend on the composition of stands and other characteristics of each forest as well as changes in general economic conditions.

A forest owner might find it advantageous to modify the resulting pattern of production for a variety of reasons, some of which arise from purely economic constraints internal to the forest enterprise. Economies of scale in forest operations mean that short-run adjustments in production above or below the most efficient capacity of harvesting will drive up costs. It will therefore be more efficient to smooth out sharp fluctuations in production, and adjust productive capacity accordingly. Correspondingly, if a manufacturing plant is dependent on the supply of timber from the forest its capacity constraints will add to the advantage of a steadier supply of timber.

The sensitivity of product markets to changes in the timber supply may also make it disadvantageous to simply harvest all stands according to the optimum rotation rule. For example, if the resulting increases in the rate of harvesting would drive down the prices obtainable in log markets and decreases would cause prices to rise, the forest owner could improve his returns by scheduling more regular production.

For all these reasons forest owners may want to modify their patterns of harvesting over time to maximize the value of production from the forest as a whole. The resulting pattern of harvesting will depend on the circumstances of each case; the only general conclusion is that it is likely to be smoother over time than would result from harvesting each stand at its independently optimum rotation age, and the smoothing effect will be greater the sharper and greater the fluctuations in harvest rates that would otherwise take place. However, it is important to note that the result would not likely be a constant harvest rate over long periods. Irregularities in the inventory and changes in costs and prices will mean efficient harvest rates will change, albeit more gradually, over time.

The adjustments in harvest rates described above can be expected to be made by forest owners. The timber supply in many countries and regions is dominated by independent forest owners responding individually to market incentives in this way. Insofar as they gain from improving the economic efficiency of production they also advance the interests of society at large.

However, forest regulation often goes further than adjustments to enhance economic efficiency. Governments often manage public forests, or regulate private forests, in such a way as to maintain harvest levels over long periods, sometimes in perpetuity. To the extent that these policies conflict with economic efficiency and producers' economic incentives, they must be imposed by regulation. The remainder of this chapter deals with these yield regulation policies.

A variety of market imperfections can distort producers' decisions and lead to undesirable rates of resource use; externalities, ignorance, imperfect competition, and divergence between private and social time preference are among those described in Chapter 2.

These have often led governments to intervene in various ways to mitigate their adverse effects. However, the dominant concern of governments in regulating harvest rates is industrial stability, arising from fears that uncontrolled producers reacting freely to swings in forest products markets will cause unstable employment and incomes. A longer-term concern is that unregulated exploitation may lead to resource depletion, eroding employment opportunities and the economic base of regional economies.

THE REGULATED FOREST

Harvest regulation or yield planning has a long history in forestry theory and practice, extending from the fourteenth century in Europe. The focus of this attention has always been on how to regulate harvests in order to reconstruct the forest inventory so that it will be capable of sustaining a yield that meets specific quantitative objectives.

Underlying many of the classical approaches to forest regulation is the concept of the *normal forest*. This is a forest with an even distribution of age classes, so that it is capable of yielding the same volume of timber every year in perpetuity.

The simplest possible case has already been described: a forest of uniform land productivity consisting of stands of equal areas, one of every age from zero to the rotation age. Each year, the crop that reaches rotation age can be harvested and, providing it is always restocked, the harvest can be maintained indefinitely at a rate equal to the growth of the whole forest. Even today, some variant of this simple model of a normal forest is the forest structure that many forest management agencies strive to achieve through harvest regulation.

Though simple in concept, a normal forest presents a complicated problem in practice. The forest to be regulated is often delineated by the boundaries of forest properties, which can range from small woodlots to huge holdings. On public lands they are more often determined with reference to geophysical features that take account of access and distance to manufacturing centres. Their size and productivity, and the rotation period chosen, will determine the structure of the forest needed to maintain a steady yield with a given level of silviculture and management effort.

Because a large tract of forest land is never uniform in productivity, a constant yield can be achieved only if the areas assigned to different age classes are varied according to the varying quality of the land. Variations in site productivity mean that the rotation period must vary as well, complicating considerably the problem of scheduling and balancing the sequence of harvests.

Once in place, a fully regulated normal forest implies a dynamic, steady-state cycle of harvests, continuously balanced by growth, which depends on specific and unchanging conditions. If a constant yield is to be maintained, no changes can be made in the land base, in rotation ages, in silvicultural effort, or in standards of utilization.

Because management planning for a normal forest inherently involves projections over long periods of future time, these rigid requirements for maintaining the constancy of growth and yield are not likely to be met. Advances in forest science and silvicultural practice can be expected to improve production over time; technological change is likely to alter the proportion of the crop that can be recovered and utilized; and changing economic circumstances will affect the costs and prices that determine the most advantageous forest management regimes. As a result, the fully regulated normal forest is rarely achieved in practice. Nevertheless, it is a theoretical model with considerable intuitive appeal to foresters and policymakers, and it provides the target to which a great deal of forest regulation is directed.

TRANSITION TO A NORMAL FOREST

In many forest regions the chief regulatory problem is not how to manage normal forests but how to create them from forests that consist of unbalanced distributions of age classes. For example, in regions of North America where much of the forest consists of original "old growth" timber, well beyond the harvesting age for managed forests, this stock must be removed and reforested in a carefully controlled pattern over time to create the needed age structure for a regulated forest. In countries such as the United Kingdom, New Zealand, and Portugal the problem is the opposite; with new and expanding plantation forests, the shortage is in the older age classes. For these and other circumstances much effort has been put into methods for transforming irregular forests into a fully regulated state.

From an economic standpoint the planned harvest schedule during the transition period is the issue of primary importance. This is because the transition phase usually extends over many decades and, as noted in earlier chapters, the force of discounting typically reduces the contribution of harvests beyond this period to a relatively small proportion of the total present value.

The two general approaches in transforming an irregular forest to one with an even age class distribution are referred to as *area control* and *volume control*. In the hypothetical case of a forest on land of uniform productivity, area control involves harvesting and reforesting each year a portion of the forest equal to the total area divided by the rotation period chosen for subsequent crops. Then, in the year that the last of the original forest is harvested, there will be an even gradation of age classes from zero to the rotation age, and thereafter the forest can be managed as a normal forest.

Area control will not produce a steady yield during the transition period, of course. The harvest level at any time will depend on the structure of the original forest and the sequence in which it is harvested. For example, if it contained old-growth forest and this was harvested first, the yield may decline when harvesting moved on to equal areas of younger stands. In contrast, a forest of young plantations might initially require cutting stands well before their planned rotation age, yielding low volumes which would subsequently increase.

In contrast, volume control aims at stabilizing harvests over the transition period, while allowing the areas harvested to vary. It involves fixing the volume to be harvested each year, referred to as the *allowable annual cut*, using some formula that takes account of the volume and age distribution of the original forest inventory and its rate of growth.

One such formula that has been widely used in the United States and Canada is the so-called *Hanzlik formula*, originally formulated to guide the conversion of virgin forests of the Pacific Northwest into normal forests in one rotation period. It specifies the allowable annual cut, aac, for one rotation period as:

$$aac = (Q_M \div A) + mai$$

where Q_M is the volume of timber in the forest that is already beyond the rotation age, A, planned for subsequent crops, and mai is the mean annual increment, or average annual growth, expected on the areas occupied by immature stands over their planned rotation. Note that if the initial forest consisted entirely of old timber that had ceased to grow, mai would be zero, and the formula would specify cutting the original stock in equal increments over A years. Conversely, if the forest consisted entirely of stands of less than the rotation age (as it would in any event at the end of the transition period) the first term on the right-hand side of the equation would drop out, indicating an allowable annual cut equal to the forest's growth rate.

Volume control of this kind, in contrast to the simple form of area control described above, can generate steady harvests from an irregular forest during its conversion to a normal forest. However, at the end of the conversion period of one rotation, both methods are likely to call for some adjustment to the long run sustainable yield of the regulated forest. As suggested earlier, if the original forest contains a predominance of old-growth stands, yielding larger volumes of timber per hectare than can be expected from subsequent crops at their rotation age, the Hanzlik formula would call for a downward adjustment, reflecting the so-called "fall down" phenomenon associated with the conversion of natural forests. If the initial forest contains mainly immature stands, however, the allowable annual cut during the transition phase must remain below the ultimate sustainable yield and growth rate.

A variety of formulae or rules can be used to smooth out adjustments to the harvest rate over time. For example, if the level of harvesting must be reduced because of the "fall down" when an old-growth forest is converted to a managed forest, periodic recalculation of the Hanzlik formula as the overmature inventory is depleted will provide an allowable annual cut that declines more gradually to the forest's sustainable yield. Alternatively the derivation of the allowable annual cut can be constrained to ensure that it does not change by more than a specified percentage over any decade or other period. Some regulatory agencies in the United States have pursued an objective of "non-declining even flow," which seeks to avoid any downward adjustment. This presents complex problems for harvest scheduling in converting old-growth forests, and usually extends the conversion over a much longer period than one rotation.

None of these decision rules for scheduling harvests takes account of their economic implications, though the results often have significant consequences for the economic efficiency of forest enterprises. The cost, or extent to which a rule of this kind impairs efficiency, can be measured as the difference between the present worth of planned harvests using the rule and the present worth of harvests under the most efficient schedule of harvests.

For any particular forest, this difference will depend on the initial forest structure and the structure to which it is to be converted. However, the potential effect can be illustrated by considering a uniform forest of 1000 hectares, originally consisting entirely of overmature old-growth timber that has ceased growing, worth $20,000 per hectare. We will assume that subsequent stands will grow in volume and value as indicated in Table 2 in the preceding chapter, and will be harvested at the optimum rotation age of fifty-eight years. For simplicity we shall assume as well that all costs and prices are constant over time and are unaffected by the rate of harvesting, and that the applicable interest rate is 5 per cent. Under these conditions the most efficient regime would involve harvesting the entire forest in the first year, which would yield $20,000 × 1000 = $20 million. To obtain the total value of the forest enterprise we must add to this the present value of the future crops; we calculated this in Chapter 7 as the site value of the bare land, of $827 per hectare. Thus the total present value of the forest becomes $20,827,000, as indicated in Table 3.

TABLE 3: Values generated under alternative harvest schedules

	Unregulated harvest schedule[a]	**Regulated harvest schedule**[b]
Harvest of original timber	$20,000,000[c]	$6,489,405[d]
Value of subsequent crops	827,000[e]	268,338[f]
TOTAL VALUE	$20,827,000	$6,757,743

NOTES

[a] Involves harvesting all the timber in the initial forest in the first year.
[b] Involves harvesting 1/58th of the initial inventory of timber in each of the 58 years of the forest conversion period.
[c] The value of 100 hectares of timber worth $20,000 per hectare.
[d] The present value of 1/58th of the value of the initial timber (1/58 x 20,000,000 = $344,826) accruing each year for 58 years, using Equation 6, Chapter 6.
[e] The site value of 1000 hectares at $827 per hectare.
[f] The present worth of the site value of 1/58th of the forest (1/58 x 1000 x 827 = $14,259) accruing each year for 58 years, using Equation 6, Ch. 6.

Because we have assumed a uniform forest, the area control and volume control methods of converting such a forest to a normal forest involve the same regime; that is, harvesting 1/58th of the initial timber each year for fifty-eight years. We can calculate the present value of these harvests using Equation 6 in Chapter 6, that is

$$V_o = \frac{344{,}826[(1.05)^{58} - 1]}{.05\,(1.05)^{58}}$$

$$= \$6{,}489{,}405$$

or roughly $6.5 million. Using the same formula the value of subsequent crops can be calculated as the present value of 1/58th of the total site value accruing each year over the same period, which amounts to $268,338. The total present value of the regulated harvest regime is thus $6,757,743, as shown in Table 3.

In this example, spreading the harvest evenly over the first rotation period reduces the potential value of the forest enterprise by nearly two-thirds of its potential value. It should be re-emphasized, however, that this result follows from the particular assumptions of this example; the impact of harvest regulation on any forest depends on the particular conditions relating to its initial inventory, the rotation period and interest rate chosen, and other conditions.

Nevertheless, regulations that spread harvests further into the future than is economically efficient substantially reduce the value of forest enterprises. Our simple example illustrates two ways in which this impact is felt. One is the reduction in value of the initial forest that results from postponing its harvest. The other is the reduction in the value of subsequent crops that results from delaying the replacement of the initial forest with new stands. A third way, not illustrated in this example, is the loss in value that may result from a regulatory regime that prevents producers from responding to market swings, constraining them to produce less than would be most profitable when prices are high and more when prices are low.

Finally, note that the extensive margin of operations for harvesting original or virgin timber is unlikely to be the same as the extensive margin for continuous forestry. Land of relatively low productivity may support old growth timber that is profitable to harvest, yet cannot generate a positive net return in growing subsequent crops. Or, even highly productive land may support valuable timber but, once it is harvested, find its highest use in agriculture, urban development, or some other non-timber use. The history of land development and use in North America offers many examples of both these circumstances. Thus the geographical base of forest operations usually will become smaller during the transition from original to managed stands. In the terms used in Chapter 5, this can be described as contraction of the extensive margin.

ALLOWABLE CUT EFFECT

Regulations that spread harvests evenly over time alter the impact of forestry investments. To illustrate this important effect, consider the possibility of improving the stand established at the beginning of the crop cycle at a cost of $1000 per hectare, which would double the growth in the example depicted in Table 2. Thus, at the harvesting age of fifty-eight years, the stand would contain 995 cubic metres more than without the treatment.

A decision-maker would normally weigh the cost of the treatment against the benefits of the increased yield to be realized fifty-eight years hence. However, if the forest is regulated according to a rule, such as the Hanzlik formula, that prescribes a constant harvest level over the whole rotation, the enhanced yield would be spread evenly over each year of the growing cycle, through an increase in the allowable annual cut of $(1/58 \times 995) = 17.16$ cubic metres per year. This is the so-called "allowable cut effect"; any improvement in growth, regardless of when its direct effect will be realized, will increase the prescribed harvest rate immediately and by the same amount each year for the number of years in the forest rotation. Conversely, any loss of inventory or growth due to fire or pests will be spread over the whole rotation, regardless of whether the lost timber was close to harvesting age.

The allowable cut effect, in turn, distorts the apparent economic returns from silvicultural investments. In our example, the initial cost of $1000 would normally be compared with the present value of the resulting increased yield of $13,187 that would be realized in fifty-eight years. Using a discount rate of 5 per cent and Equation 2 in Chapter 6 the benefit is

$$V_o = \$13{,}187 \div (1.05)^{58}$$
$$= \$778$$

so the project is unattractive. However, using the even-flow formula, 1/58th of this improvement would be taken each year beginning immediately which using Equation 6 in Chapter 6 has a present value of

$$V_o = \frac{(13{,}187 \div 58)[(1.05)^{58} - 1]}{.05\,(1.05)^{58}}$$
$$= \$4{,}546$$

indicating benefits exceeding costs by more than fourfold.

The allowable cut effect thus exaggerates the returns to investments in forestry, and conversely reduces economic costs of inventory losses. The magnitude of the distortion depends heavily on the structure of the forest inventory and the rotation age. For forests with a preponderance of virgin timber the apparent returns to silvicultural investments that enhance growth in new stands can be astronomical, because the allowable cut effect translates the improved growth into an immediate increase in the harvest of mature timber. But this result has less to do with the real impact of the silviculture than with the artificial effect of the yield regulation.

Harvest regulation thus can have a major impact on the financial incentives of forest managers. An important question therefore is whether the allowable cut effect should be incorporated into economic evaluations of investment opportunities in silviculture, protection, or additions and deletions of land in the regulated forest. The answer must be that, notwithstanding the distortions, it should guide the evaluation as long as the decision-maker is constrained to manage the forest within such a regulatory regime. If the regulatory policy will govern the actual outcome of the investment, it must be recognized in the evaluation. However, the allowable cut effect is an unwarranted distortion of the returns on a forestry investment when the decision-maker can use normal investment criteria and behave accordingly. And, of course, any evaluation of the regulatory policy itself must take full account of the extent to which it influences behaviour and reduces the efficiency of forest investments.

BENEFITS OF SUSTAINED YIELD

To evaluate harvest regulation policies we must weigh the costs against the benefits. The extensive literature on harvest regulation suggests a considerable variety of benefits that may flow from such regulation, loosely termed sustained yield policies. Before turning to the implications for industrial and social stability we should note certain other, more tenuous, arguments in support of these policies for regulating harvests at more-or-less constant levels.

National security. Historically, an important motive of governments in promoting stable forestry was strategic. In pre-industrial Europe, wood provided the primary source of energy and fuel for both domestic and commercial needs. It was also the primary raw material in construction and in shipbuilding. Indeed, a good deal of early forestry was motivated by naval requirements. Because wood was such a critical raw material, dependence on foreign supplies was risky; and because it was so bulky and costly to transport, regional economies needed local supplies. So sustained yield policies were part of national strategies for defence and economic security.

Economic and technical change has made these considerations much less relevant today. Wood is no longer the critical source of energy, or raw material for military and industrial purposes, and developments in transportation have further eliminated dependence on local timber supplies. In any event, strategic considerations suggest a need to maintain stocks of timber that can be called on in case of emergency, but not necessarily sustained yields either before or during such events.

Moral obligation. Underlying much discussion about resource management policy is a conviction that today's managers have a moral obligation to future generations to ensure that forests are passed on in an enhanced, or at least unimpaired state, and that this duty can best be fulfilled through a sustained yield policy.

The matter of moral obligation raises both a technical issue and a political issue. The technical issue is the effect of regulating yields on a forest's productivity. It has already been noted above that yield regulation is likely to diminish the productivity of a forest in economic terms. Moreover, the future productive capacity of a forest can be increased in terms of value or quantity of timber by building up the stock in the present, which implies abstaining form harvesting rather than maintaining a steady harvest. Or, if productivity is measured by the growth rate it can be increased by means of protection, silviculture, and other management measures. In short, conversion of the forest to an even gradation of age classes and the maintenance of harvest levels are not essential for protecting the future productivity of forests.

The political issue is whether we have an obligation to leave our forests to our successors in as productive a state as we found them. To the extent that this proposition is a purely ethical or political view, it is not susceptible to economic analysis. However, it does not follow directly from a concern to advance the economic welfare of future generations. This is because the wealth of future generations, and the productivity of their economy, may be enhanced by depleting some forests and converting them to other productive assets, expanding them to take advantage of new opportunities such as genetic improvements, or transferring some land to other uses. Historically, we have responded in all these ways, but they are not consistent with maintaining forests in a constant state.

Nevertheless, there is a widely held view, rooted in the conservation ethic and currently reflected in the search for "sustainable development," that we are temporary stewards of renewable resources. There is also a widespread perception, based on "cut-and-run" forest operations early in this century, that the forest industry, left to itself, will not adequately consider the interests of future generations.

Whether the historical rate of timber harvesting was too rapid is debatable but, if governments accept a duty to intervene, the analytical problem is to identify how the unregulated pattern of resource use diverges from the interests of future generations, then to evaluate means of correcting the distortion. The problem and the most effective solution can be expected to vary at different times and places. They may call for measures to ensure the preservation of the soil and its fertility, its prompt regeneration after harvesting, slower rates of harvesting or preservation of the forest environment. But because the threat of unregulated activity varies, the stewardship argument does not lead to a single policy prescription for all circumstances. In particular it does not imply a specific form of harvest regulation.

Multiple use. It is sometimes suggested that non-timber benefits of forests, such as environmental and recreational values, livestock forage, water and aesthetic benefits, can be protected or enhanced through sustained yield management. On careful examination this, too, is a tenuous argument. As noted in Chapter 4, the most favourable management regime for any particular benefit depends on the ecological and other circumstances of each forest, and different forest products and services call for different management regimes. Aesthetic values may be improved by maintaining a heavier cover of timber; wildlife and livestock might benefit from less, and so on.

In general, it is unlikely that an even gradation of age classes, and a constant harvest rate would maximize non-timber benefits of any particular kind, or any particular combination of them. Most of these values are affected much more by such things as how much natural forest is preserved, how roads systems are designed, and how logging is conducted than by the regularity of harvests over time.

Other arguments. A variety of other arguments have been advanced in support of sustained yield policies, such as silvicultural improvement, risk reduction, and protection of forest owners against their own ignorance or shortsightedness. While one or another of these may apply in particular circumstances, none offers a logical rationale for a general policy of regulated sustained yield.

Economic stability. By far the dominant rationale for sustained yield policies is regional economic and industrial stability. The argument is that a steady harvest of timber will stabilize harvesting and wood manufacturing industries and hence also regional employment and income. This takes us into macroeconomics, the branch of economics that deals with the general level of economic activity, prices and employment, the causes of instability and growth, and the means of controlling these phenomena. Despite the prevalence of sustained yield policies, the macroeconomic significance of the forestry sector is not well developed in forestry literature. Here it is important to identify the linkages between the level of timber harvesting in a region and the level of regional economic activity.

First, insofar as sustained yield policies are regarded as governmental means of promoting stability, it is important to clarify the economic variables that they are intended to stabilize. In particular, the concern for long-term economic stability must be distinguished from the problem of short-term instability. Forest products markets suffer from short-term cyclical fluctuations mainly because of shifts in demand (in contrast to markets for agricultural products, for example, which fluctuate mainly because of shifts in supply). These demand shifts are usually due to forces that have little to do with forest policy; often the cause may be as remote as interest rate changes or swings in international markets. Sustained yield, as such, deals only with the supply side of timber markets.

Moreover, stabilization of supply in the face of fluctuating demand cannot stabilize prices. In the absence of controls, producers reduce production when prices fall and expand when prices rise, cushioning the pressure on prices both ways. A requirement to maintain production at a constant level in the face of shifting demand would aggravate the resulting swings in prices. Pressures on inventories and financial performance would be increased as well. Thus sustained yield policies normally allow for some adjustments in harvest levels in response to short-run market swings, focusing instead on the stability of economic activity in the long term.

Second, the impact of the forest sector on the level of regional economic activity depends on its importance in the regional economy. Except in rare circumstances, the forest industry is only one among many comprising a regional economy. Each sector follows somewhat different long-term trends in growth and employment, and macroeconomic policy is usually concerned with stability of incomes and employment in the aggregate. The impact on the stability of regional economies of regulating the forestry sector, independently of the other sectors in the region, will vary widely.

Third, harvest regulation does not stabilize the forest industry but rather the flow of raw material it uses. This distinction is important because stability in regional economies is usually sought in terms of employment and income. Over the long periods of time considered in forest regulation, the relationship between the quantity of timber harvested and processed and the amount of labour employed is likely to change significantly. Technological advances and substitution of capital for labour tend to increase labour productivity by a few percentage points per year, so that over a period as long as a forest rotation much less labour will be employed per unit of timber produced. Accordingly, stabilizing the production of timber over long periods will not ensure a stable employment level.

With respect to regional income, payments to labour are usually most relevant, because payments to capital are typically smaller and often accrue outside the region. Labour income is largely a product of the number employed and the average wage rate, which often follow opposing trends. Advancing productivity is associated with declining employment per unit of output while wages rise. Both technological change and trends in real wage levels are governed by forces in the broader economy, and how they will balance out over long periods and affect the level of regional incomes in the forest sector will vary among regions and over time, to a large extent unpredictably.

Other changes also alter the long-term relationship between the volume of timber harvested and regional employment and income. Changes in products and manufacturing processes are likely to occur. New transportation systems are likely to alter the links between particular forests and the manufacturing centres they supply. And the organization and structure of the industry are likely to shift the geographic distribution of activity.

For all these reasons it is difficult to draw many general conclusions about the relationship between the supply of timber in a region, or from a particular forest, and the long-term stability of the regional economy. Some writers have argued that there is little evidence to support the proposition that an even flow of timber over long periods will promote regional stability, and that it is likely, instead, to retard growth, adaptation to change, and reallocation of resources.

In any event it is clear that a policy aimed at stabilizing long-term regional economic activity must look beyond the raw material supply for a particular industry, and take account of the rest of the regional economy and the particular circumstances and trends in other sectors. It follows that traditional concepts of the normal forest and sustained yield are not likely to be sufficient to ensure that the forest sector will make its most effective contribution to the stability of regional economies.

NEW APPROACHES TO FOREST REGULATION

Chapter 1 emphasized the importance of social objectives in designing public policies and assessing performance. The objectives of private and public forest owners vary, as do the circumstances in which they operate, and today they are undoubtedly different and more complicated than those that led to the development of the traditional concepts of sustained yield forestry many years ago.

In recent years much more sophisticated approaches to the question of the optimum rate of use of resources over time have been developed. Theoretical solutions under various conditions and constraints have been worked out, often using optimal control theory, though the mathematics are formidable. However, advancing computer technology has provided practical means for rigorous analysis of complicated systems and, in response to the growing need for careful forest management planning, these new tools have been adapted to assist in timber harvest scheduling.

Elaborate computer-based models, capable of incorporating enormous quantities of information about a forest and the economic conditions of timber production, are now available to assist planners in evaluating alternative strategies of management much more comprehensively than was previously possible. These models enable analysts to assess the implications of all possible harvesting schedules in terms of a variety of objectives, making obsolete the traditional strategy of steady volumetric yield.

One of the many advantages of these computer-based analytical systems is that they can build in economic assessments of the alternative possible forest management regimes. Moreover, the otherwise complicated and laborious computations of economic implications can be done easily and quickly, to assist decision-makers in their search for plans that will best serve their objectives.

These modelling techniques have become practical only with the advent of advanced computer technology, capable of incorporating and analysing the volumes of relevant data on forest growth, responses to silviculture, interactions among competing forest products and services, and the costs and prices that bear on forest management planning. As explained in Chapter 11, a growing variety of such models are becoming available to assist forest managers in analysing harvest schedules and other forest management programs. They are rapidly replacing forest managers' reliance on traditional concepts of sustained yield and in the process providing much more rigorous economic support for management decisions.

REVIEW QUESTIONS

1 How does the supply of timber from a particular forest affect the market supply of timber as described in Chapter 3 when (a) the forest is only one of many sources of supply for the market, and (b) when the forest is the only source of supply for the market.
2 Why is it often necessary to modify the harvesting regime that maximizes the return from an individual hectare of forest in order to maximize the return from the forest as a whole?
3 What is a "normal forest"? What silvicultural conditions must be met for a normal forest to yield a constant harvest over time?
4 Use the Hanzlik formula to calculate the allowable annual cut for a forest of 5000 hectares of uniform productivity, half of which is occupied by old growth timber of 2000 cubic metres per hectare, the other half by twenty-year-old second growth having a mean annual increment of fifteen cubic metres per hectare over the planned rotation period of fifty years. How does your answer compare with the allowable cut after the old growth is depleted?
5 Calculate the present value of the old growth in question 4 assuming it will be harvested in equal annual amounts over the next fifty years. Use a discount rate of 4 per cent and assume that all timber is valued at $20 per cubic metre. How does this compare with the value of the timber if it were all harvested immediately?
6 Why will a stable level of forest harvesting sometimes fail to ensure the economic stability of nearby communities?

FURTHER READING

Bell, Enoch, Roger Fight, and Robert Randall. 1975. ACE: the two-edged sword. *Journal of Forestry* 73(10):642–43

Berck, Peter. 1979. The economics of timber: a renewable resource in the long run. *Bell Journal of Economics* 10(2):447–62

Bowes, Michael D., and John V. Krutilla. 1989. *Multiple-Use Management: The Economics of Public Forestlands*. Washington, D.C.: Resources for the Future. Chapter 4

Buongiorno, J. and J.K. Gillees. 1987. *Forest Management and Economics*. New York: Macmillan

Davis, L.S., and K.N. Johnson. 1987. *Forest Management*. 3rd ed. New York: McGraw-Hill. Chapters 13, 14, and 16

Grayson, A.J., and D.R. Johnston. 1970. The economics of yield planning. *International Review of Forestry Research* 3:69–122

Heaps, Terry, and Philip A. Neher. 1979. The economics of forestry when the rate of harvest is constrained. *Journal of Environmental Economics and Management* 6:297–319

Hyde, William F. 1980. *Timber Supply, Land Allocation, and Economic Efficiency*. Baltimore: Johns Hopkins University Press for Resources for the Future. Chapter 2

Klemperer, W. David, John F. Thurmes, and Richard G. Oderwald. 1987. Simulating economically optimal timber-management regimes: identifying effects of cultural practices on loblolly pine. *Journal of Forestry* 85(3):20–23

Nautiyal, J.C. and P.H. Pearse. 1967. Optimizing the conversion to sustained yield: a programming solution. *Forest Science* 13(2): 131–39

Thompson, Emmett F. 1966. Traditional forest regulation model: an economic critique. *Journal of Forestry* 64(11):750–52. (Also in Fay Rumsey and W.A. Duerr (eds.) 1975. *Social Sciences in Forestry: A Book of Readings*. Philadelphia: W.B. Saunders. Pp. 254–59)

CHAPTER NINE

Property Rights and Forest Tenure Systems

Out in the Forest . . .

Like many other companies, Peavey Forest Products Limited cuts timber on public lands as well as on its own private lands. Operations on public lands are carried out under a timber sale licence issued by the Forest Service on behalf of the government. The licence gives the company the right to harvest the timber in a prescribed area within three years, subject to a variety of regulations and payment of stumpage fees on the logs recovered.

The company's interest in the public forest covered by its licence is different from its interest in its private lands. Because its rights under the licence extend for only a short period, the company has no financial interest in harvests beyond that time or in providing for subsequent crops. Since the licence conveys only a right to the timber, the company similarly has no financial interest in protecting or enhancing other forest values such as water and wildlife, which are affected by logging. And because the Forest Service must be paid a stumpage fee on all logs harvested under the licence, the company's incentive to recover marginal logs is blunted. These and other differences in the character of the property rights under which it operates thus induce the company to use and manage its licensed public land differently from its freehold land.

The narrow financial interest of companies holding short-term timber sale licences explains many of the controls and regulations imposed by the Forest Service under this form of tenure. In effect, regulations are designed to ensure that the limited rights of Peavey Forest Products Limited and others who harvest public timber do not cause neglect or short-sighted use of the forest.

Chapter 1 noted the basic policy questions about forest resources. Who will own them; who will utilize them; who will manage them; and who will get the economic benefits from them are among the fundamental issues that must be dealt with through forest policy. In large part these questions are resolved through forest tenure systems, which govern the rights of owners, users, and others over forest land and timber.

In western countries these rights take a variety of forms. Some forests are owned as private property, subject to varying degrees of governmental regulation. Some are held by governments, referred to in Canada as federal or provincial Crown land and in the United States as federal or state public lands. But timber companies and others often hold rights over land and forests owned by governments, through various forms of leases, licences, and permits. The simple distinction commonly made between private and public property is misleading in this area; in fact, property rights in forest land and timber take many forms. Various types of tenure, which refers to the holding of property rights, fix the rights and responsibilities of tenants, landlords, and governments, and thus are important instruments of forest policy.

In discussing the conditions for efficient use of resources in a market economy in Chapter 2, we noted the importance of the producer's control over his inputs. We also noted that producers of forest products rarely enjoy complete control over the forest land and timber they use. The relationship between the rights held by users of forest resources and the efficiency with which forests are developed and used is the subject of the present chapter.

PROPERTY, VALUE, AND ECONOMIC EFFICIENCY

Property rights define the extent to which the holder of the rights can enjoy the benefits of particular goods or assets. We commonly think of property in terms of a tangible good, like a bicycle, a house with its plot of land, or a forest. Someone owns the thing, so it is his property. But in law, property is conceived differently; it is a defined set of rights that the holder has over something. Lawyers are taught to think of someone's property in something as a bundle of sticks. Each stick represents a right, like the right to exclude other people from using the thing, the right to enjoy the economic benefits it conveys, the right to sell it to someone else, and so on. The bundle may be big or small, reflecting the range of possible rights involved.

For present purposes, we will adopt the lawyers' conception of property, emphasizing rights rather than things. Property is a major branch of legal studies, supported by very extensive literature and long tradition. Here, we only need to draw attention to the variety of property rights in forestland and timber, and their economic implications.

The value of any property depends upon two factors. One is the inherent physical and economic properties of the resources or goods over which the property rights extend, which govern the benefits they can yield. The other is the extent to which the property rights enable the holder to enjoy these attributes. Property rights over even very valuable resources will not be worth much if the rights themselves are highly restricted or truncated. If the rights extend for only a short period, if the holder is restricted from selling the rights, or if his rights allow others to share the benefits, the value of the property will be correspondingly lower.

Thus the value of property rights over a forest depends, first, on the inherent productivity of the forest in terms of the value of forest products and services it can yield and, second, on how much of these benefits the rights allow their holder to enjoy. If they permit him to use the forest forever the rights will be worth more than if he can use it only for a year or two; if they include the right to use the water, minerals, and agricultural values they will be worth more than if they convey only the right to use the timber; if it is not taxed it will be worth more than if its benefits must be shared with the government, and so on.

Economists put great importance on property because it governs the efficiency of resource use throughout an economy as well as the distribution of income. In the theory of the pure, perfectly competitive market system all factors of production are owned privately and property rights are "complete" in the sense that they are not restricted or qualified. The bundle of sticks is big. With unrestricted rights to do whatever they want with their assets, owners are consistently driven by incentives to put them to their highest and most rewarding use, thereby contributing to social welfare. Through this same process, ownership of productive factors determines the distribution of income. Economists of the property rights school have argued that comprehensive and "complete" property rights, precisely defined, in all factors of production would insure maximum possible efficiency in economic production through market mechanisms.

EVOLUTION OF FOREST PROPERTY

Theories about the origins of property suggest that private rights emerge from an original regime with no property, or common property, when resources become scarce and valuable. Then, if users lack means of defining and establishing their claims to land and natural resources, and protecting them from others, they begin to interfere with each other's production and cause other costly inefficiencies. Sooner or later the potential gain from eliminating this interference exceeds the cost of organizing exclusive private property rights to eliminate it. In short, as long as resource values are low, the benefits from organizing property do not justify the cost, and the system of users' rights remains crude. But as resource values rise, raising as well the potential gain from improved allocation arrangements, more sophisticated systems of property rights can be expected to emerge.

The evolution of property rights over natural resources in North America is consistent with this model. Early settlers helped themselves to the abundance of fish, timber, water, and other resources of the land. There was no scarcity in the economic sense and no allocation problem. So there was no need to worry about the complications and cost of organizing property rights.

Gradually, with settlement, pressures developed and so did the need to allocate resources. But pressures on resources grew unevenly, of course. The need arose to parcel out land in settlements for living space, and around them for agriculture, but not beyond the "frontier" which, by definition, was unappropriated. There was less urgency, in pioneer days, to worry about allocating rights over water, fish, and timber. But as the frontier receded and development proceeded with it, one after another natural resource needed to be allocated among competing uses and users.

Thus we find, in North America, well-developed systems of property over resources that became valuable long ago, such as urban and agricultural land and minerals, but rights to water, fish, and wildlife remain relatively primitive in many jurisdictions.

In the early British colonies, the forms of property in land were derived from English common law. Frequently, the land was acquired from native occupants by the Crown through treaty or conquest, and the Crown then granted title to settlers, land development companies, railroads, and other private interests. Early grants usually carried with them the full range of rights under traditional common law, including unrestricted rights to use the surface of the land, whatever lay beneath the surface, the water and timber on it, the wildlife, and so on. In the early days, grants of title were the simplest, the least expensive, and so the normal way of obtaining rights to resources.

Later, mainly in the decades around the turn of the century, governments throughout North America turned away from the policy of granting complete or outright title to land and resources except for land needed for urban development and agricultural purposes. For other resources, notably timber, governments designed rights in the form of leases, licences, and permits, which they could issue to private parties to allow them to use specified resources while the government continued to hold the title to the land. Some of the early leases over timberland were much like outright private ownership insofar as they carried long terms and provided their holders with the exclusive use and benefit of the resources with few restrictions and controls. Others, like timber sales, authorized the use of a specific amount of timber, for a specific purpose, within a short period. The result is a spectrum of property rights in forestland and timber, ranging in duration, comprehensiveness, exclusiveness, and other characteristics which have important economic implications, discussed below.

Governments have also developed regulatory means of reconciling competing uses and users of public resources, to some extent as an alternative to systems of property. The cost of organizing and establishing property rights over some resources, such as those that yield general environmental benefits, can be exceedingly high, and so governments have developed ways of regulating their use instead. Examples include pollution controls, and recreational fishing and hunting regulations. So today, we find a complicated mixture of public, private, and common property, and policies that rely on a mixture of market processes and governmental regulation.

DIMENSIONS OF PROPERTY AND THEIR ECONOMIC IMPLICATIONS

Property has several dimensions of economic importance; the main ones for us to consider here are the following:

Comprehensiveness

Comprehensiveness refers to the extent to which the holder of the property has rights to the full range of benefits from an asset. For example, when someone holds a tract of forestland under a private freehold he can usually claim the full range of values generated from timber, agriculture, recreation, water, and so on. If his rights are in the form of a timber sale licence, however, he is usually restricted to the benefits of timber production alone.

In this, and in other dimensions of property, there is a spectrum of possibilities. A particular form of property right occupies a particular place on the spectrum.

The degree of comprehensiveness of the user's property rights has important implications for the economic efficiency with which he will manage and use a forest. If someone has fully comprehensive property rights over a forest, he can be expected to maximize the value generated by all its attributes and possible uses, compromising one in favour of another whenever it is advantageous to do so. In contrast, if someone holds the rights to the timber in a forest, but not to the water, wildlife, or other benefits which are affected by his timber operations, and if he does not need to compensate anyone for any adverse effects on these other values, he will tend to ignore them. In these circumstances the holder will seek to maximize the benefits which he can claim, disregarding those which he cannot claim and any external costs or benefits he may inflict on others. When users thus fail to take account of all the effects of their decisions, the aggregate benefit from all the values will fall short of its potential. Governmental regulation is the only means of overcoming these impediments to socially desirable patterns of resource use.

Such problems are common in forestry. Companies exercising their rights to cut timber impinge on the benefits of those who have rights to recreation, aesthetic enjoyment, or wildlife in the same forest. In these circumstances no single decision-maker has an incentive to search for the optimum combination of uses, as described in Chapter 5.

Sometimes rights are fragmented among private parties, who separately hold rights to water, timber minerals, and so on. Sometimes they are fragmented among governments, as when a provincial, state, or local government claims rights over the timber, while a federal government has authority over fish or wildlife habitat. This often results in serious conflicts among governments and various private holders of resource rights.

It is important to note that these conflicting interests are created by artificially separating rights to the same resources. If they were all held by one party the conflict would not arise; he would determine the most advantageous balance of uses. Or, even if the rights to the different resources or attributes of a forest were held by different people, market processes might still produce an optimal result as long as the rights were freely tradeable. For example, if the holder of timber harvesting rights threatened the interests of the holder of water rights, and the threatened water values exceeded the benefits of logging, the holder of the water rights could simply buy out the logging rights to the advantage of both and of society as a whole. This sort of beneficial transaction is commonplace where rights are private property. But such market processes cannot be relied upon when property rights are not well defined and held by someone who can transact in them. Then misallocations are likely and conflicts often remain unsettled unless governmental regulation is invoked to resolve them.

Duration

Duration refers to the length of time over which the property rights extend. Private freehold covers rights in perpetuity, while leases and licences normally have a finite term.

The duration of property rights is important because it determines the extent to which the holder will take account of the future impact of his actions. If the rights over a forest extend for a long period, the holder can be expected to consider carefully the relative economic advantage of harvesting now or in the future, the returns to investments in silviculture, and so on. But if his rights terminate shortly, he will disregard such long-term considerations. (In the framework of Chapter 6, the effect is tantamount to his discount rate rising to an infinitely high value beyond the date at which his rights expire.)

Because of the extraordinarily long-term considerations involved in forest development and silviculture, the duration of rights over forest land is a special problem. Unless their rights extend over the several decades it takes to grow forest crops, those who harvest timber will lack adequate incentives to provide for reforestation. Then subsidies or regulations might have to be devised to ensure appropriate investments in resource management.

In addition, the duration of property rights over forests is often a primary determinant of the holders' security of supply of timber. Security of raw material supply, in turn, is a major influence on decisions about investment in manufacturing and other facilities, and hence on the efficiency of resource use.

Benefits Conferred

Another important dimension of property is the extent to which it provides its holder with a right to enjoy the potential economic benefits from an asset such as a forest. This is often constrained by governmental restrictions on how the forest can be harvested, managed, or utilized. Regulations that restrict the rate at which timber may be harvested, require loggers to recover uneconomic logs, impose measures to protect environmental quality, prohibit exports of unmanufactured timber, and direct some of the return into the public purse through taxes, royalties, and other charges all impinge on the benefits that flow to the holders of forest property.

Restrictions on the extent to which a holder of rights over a forest can enjoy its potential benefits obviously affects the value of his forest property and hence the distribution of income. Moreover, they usually create incentives to alter the way resources are used and so affect efficiency as well.

Transferability

The transferability of property refers to its capability of being bought and sold or assigned to someone else. Transfers of forest property are sometimes restricted. For example, the terms of temporary licences and leases often restrict the licensees from transferring their rights to someone else, or require them to obtain the consent of the governmental or private licenser to do so.

If property is absolutely non-transferable, it has no market value. The only way its holder can benefit from it is by exercising the rights himself. This restriction obviously affects the distribution of income and wealth and it also impedes efficient allocation. Economic efficiency depends upon the acquisition of resources by those who can generate the most value from them, who are thereby able to offer more and bid them away from less efficient users in a competitive market economy. Impediments to the marketability of assets prevent them from being transferred to those who can use them most productively.

A related issue is the divisibility of property. To take full advantage of economies of scale and changing economic opportunities, entrepreneurs must be free to divide and combine rights to resources. This is sometimes restricted in forest property by governmental prohibitions on the subdivision of rights such as leases and licences on public land, or regulations about maximum and minimum allocations under particular forms of rights.

Exclusiveness

Exclusiveness refers to the extent to which the holder of the property can claim sole rights to the exclusion of others. The ability to exclude "third parties" is a fundamental element of property and has important economic implications.

When rights are not exclusive, and their holders compete with others for the same benefits, such as the timber in a forest, they are likely to exploit it inefficiently and too fast. Moreover, users' incentives to conserve for the future, and invest in future yields, will be weak because they cannot expect to capture the full benefits of their individual actions.

We return to this important dimension of property below.

EXCLUSIVENESS IN FOREST TENURES

These five dimensions of property rights—comprehensiveness, duration, benefits conferred, transferability, and exclusiveness—are the most important economic influences of forest property rights and tenure systems. Together they govern the extent to which producers can obtain control over forest assets, identified in Chapter 2 as a primary condition for economic efficiency in a market system.

Each of these characteristics varies across a spectrum. Consider the dimension of exclusiveness. At one extreme, the holder of rights over the forest has entirely *exclusive* rights; that is, he can exclude all other users. The traditional freehold ownership provides private landowners with exclusive rights to use the land and its attributes, such as timber.

At the other extreme, no person holds any special rights and no one can exclude anyone else. The best example of *no property* is the high sea, where no one, nor any nation, can assert legal rights over another. There is no forestry counterpart to the high sea because all forests are claimed by national governments, at least. But there are many examples of public forests to which all citizens have free and equal access for certain purposes such as recreation. Between the extremes of no property and completely exclusive property rights there is a wide range of possibilities, as illustrated in Figure 22.

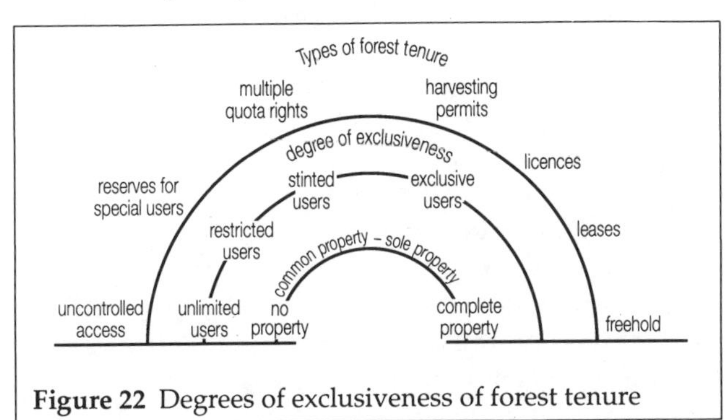

Figure 22 Degrees of exclusiveness of forest tenure

Several forms of *common property* exist. In frontier times it was not uncommon for forests to be accessible to anyone who wanted to cut timber. Such unregulated exploitation of timber is now rare, but other forest products such as game are commonly available to unlimited numbers of users. An interesting example of timber managed in this way is the ancient anomaly of the coastal strip of Newfoundland, which resulted from the early influence of fisheries traditions. Centuries ago, it became internationally accepted that coastal states had the exclusive right to fish within three miles of shore, and in western countries the fisheries were open to all citizens. In Newfoundland a corresponding strip of the forest extending three miles inland was kept available for all residents to exploit as they wanted. The effect on forestry was similar to that in open access fisheries everywhere: over-exploitation of the resources; absence of incentives on the part of users to conserve or invest in them; overcapacity in utilization; and dissipation of economic rents.

Somewhat less vulnerable to this sort of economic waste are common property resources accessible only to a restricted number of users. In North America this has become the most common regime for commercial fisheries; fishermen are required to hold licences and the number of licences is limited, but they compete with each other for undefined shares of the available resources. Today, forestry examples are mainly limited to rights issued to users of minor products such as fuel wood and Christmas trees.

In other cases common property resources are exploited by holders of licences that authorize them not only to use the resources but to take a specified quantity. The available harvest is thus *stinted*, giving the users more well-defined rights. Water rights, grazing rights, and fishing rights often entitle their holders to take a specific quantity of the resource, used in common with others. Timber harvesting rights sometimes take this form, where several users are authorized to harvest quotas of timber in a public forest without exclusive rights to any defined tract.

Most common in forestry are various forms of *sole property*, where specific resources are reserved to a single user. The most limited property rights of this kind usually take the form of permits which authorize the holder to use specific resources in very restricted ways or for particular purposes. Licences usually convey broader rights, and leases may provide rights close to those enjoyed by freehold owners, as indicated on the right side of the spectrum in Figure 22.

There is thus a spectrum of possibilities with respect to this critical dimension of exclusiveness. The tenure types mentioned here are not distinct; they overlap and merge in subtle and complicated ways.

A corresponding spectrum of possibilities exists for each of the other dimensions of property noted earlier. Comprehensiveness can range from a narrowly defined rights to rights that include all attributes and uses of the land and resources; duration from a brief period to perpetuity; benefits conferred from zero to all of the potential resource rents; and transferability from completely non-transferable to complete freedom to transfer, divide, and combine rights.

Figure 23 illustrates how any property right is a combination of these characteristics, each defined in terms of degree. Each ray in the diagram refers to a particular characteristic, such as comprehensiveness or duration, and the degree of the characteristic increases from zero at the origin to the maximum possible at its outer end. Thus the circle joining the outer ends illustrates "complete" property. The dotted lines describe a combination of characteristics which are truncated in varying degrees. An infinite variety of such combinations is possible, producing a correspondingly infinite variety of possible forms of property. However, as a result of centuries of evolution and development of the law of property, most property rights over forestland and timber now fall into a few broad categories.

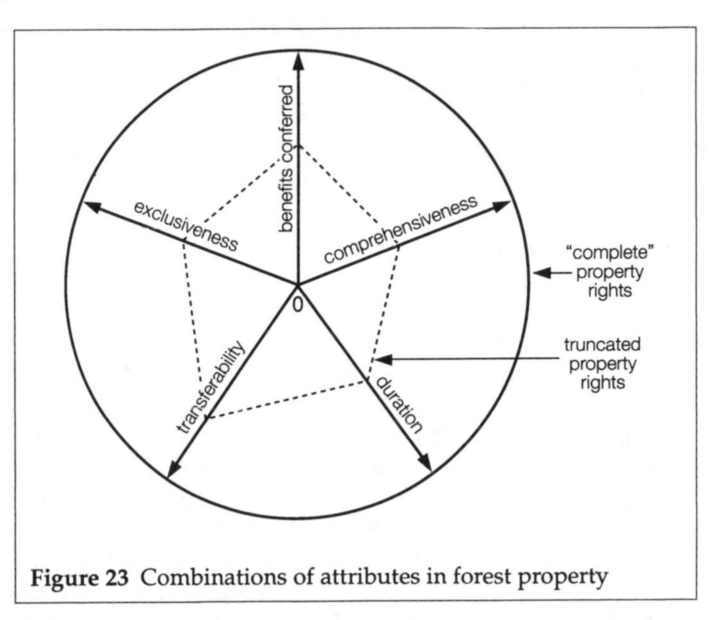

Figure 23 Combinations of attributes in forest property

COMMON FORMS OF FOREST TENURE

In North America, property rights in forest land and timber are commonly divided into two general categories; *freehold*, or what is commonly referred to as private title, fee-simple, or deeded land, and *usufructory rights*, or rights to use resources owned by others. In Canada and the United States, the latter usually take the form of a licence or lease over public or Crown land (in contrast to Britain and Europe where it is often over lands of private landowners, traditionally feudal landlords). The forms of property rights we find in North America today, used by both governments and private parties, are adaptations of common law systems evolved over countries of private negotiations, litigation, and court decisions over conflicting rights and claims.

Typical forms of rights to timber are listed in Table 4, with their characteristics in terms of the five dimensions of property discussed earlier. The forms listed are only representative; there are many other varieties of forest tenure that do not fit easily into these common types, and there is considerable variation within these types. For example, the traditional *freehold* under English common law was comprehensive, carrying with it the rights to all the attributes of the land, including the minerals and other resources under it, the water flowing over it, the fish and wildlife on it, and so on. However, over the last century most governments granting title to public lands in North America passed legislation that progressively stripped away many of these rights, reserving for governments themselves the rights to such things as subsurface minerals, water, and wildlife when they made new grants. As a result, the comprehensiveness of freehold tenure over land granted long ago is typically more complete than that over lands granted more recently. The title to lands granted in the more recently settled western states and provinces often excludes everything but the right to use the surface of the land.

Where other attributes of the land such as minerals and water are excluded from the landowner's property rights, governments often allocate the right to use them to other private parties under some form of usufructory right. However, with few exceptions, trees retain their traditional status under common law as part of the land, so title to land carries with it the title to any forest on the land.

Another significant change adopted by some governments granting title to lands was to reserve a financial interest in the forest by requiring the owner to pay a royalty, or fee, when he cut the timber. To this extent, the benefits conferred have been narrowed. Property taxes, zoning, and land use regulations, all of which have burgeoned in recent years, similarly impinge on the owner's right to enjoy the potential economic benefits from his land and forest.

An important variant of freehold tenure in some jurisdictions is

TABLE 4: Typical forms of forest tenure

Tenure	Comprehensiveness	Duration	Benefit conferred	Transferability	Exclusiveness
Freehold	complete	perpetual	all	unrestricted	exclusive
Timber lease	most attributes	long-term	most timber benefits	few restrictions	exclusive
Forest management agreement	timber harvest and forest management	long-term renewable	most timber benefits	some restrictions	exclusive
Woodlot licence	timber harvest and forest management	long-term renewable	most timber benefits	usually restricted	exclusive
Timber licence	timber harvest only	short-term	share in timber benefits	usually restricted	exclusive
Cutting permit	timber harvest only	short-term	most timber benefits	usually restricted	exclusive or non-exclusive

held by owners who contract with the government to dedicate their lands to continuous forest production. Typically, in return for an undertaking to manage the lands according to a sustained yield plan and other provisions approved by the government, the owner receives a property tax concession. This indirect subsidy reduces the degree to which taxes encroach on the benefits enjoyed by the holder, but his economic benefits are reduced to the extent that he must forego more advantageous ways of using or developing his forestland.

Early usufructory rights were typically in the form of *timber leases*. These often carried very long terms and some persist today. In some cases they are almost as comprehensive as freehold, carrying rights to other resources as well as timber. However, they typically require the holder to pay an annual fee or rent, and a royalty or similar charge on timber harvested, reducing the benefits that accrue to him. Often the terms of a lease restrict its transferability to the extent of requiring its holder to obtain the approval of the government or private landlord before selling it to someone else.

The term *forest management agreement* is used here to describe a long-term licence granted over an extensive tract of public forest land to be managed by its holder as a coherent sustained yield forest. This form of licence is now common in Canada's major timber-producing provinces, but it is rare in the United States. In some cases these licences integrate freehold lands held by the licensee with public lands under a single management plan. Under detailed provisions they assign to the licensee not only the right to cut timber but also extensive responsibilities for forest development, protection, and management under approved plans.

Forest management agreements, sometimes called tree farm licences, forest management licences, or concessions, generally carry long terms of twenty years or more and are renewable. As Table 4 suggests, the rights of the holder are confined to timber, his benefits are constrained by annual fees and levies on timber harvested, and the transferability of his rights is usually qualified or restricted in some degree.

The *woodlot licence* has a well-established, but relatively modest, place in the forest tenure policies of many jurisdictions in North America, conveying rights to very small tracts of public forests. They are not usually intended to support industrial operations, but are designed to complement farms by providing sources of fuel, fencing, and other building materials or seasonal employment in timber production. They usually convey long-term rights to the timber only, in return for modest charges and general management obligations.

Timber licence refers to a large category of usufructory rights to timber, especially on public lands. These are short-term licences to harvest defined, relatively small, tracts of timber. The most well-known form is the traditional "timber sale," but there are many variations. The licensee's rights are typically limited to harvesting the timber in an approved manner, in return for which he usually must pay a licence fee, an annual rent, and a charge against timber harvested. The licence often imposes on him a variety of other obligations as well, relating to such things as road building, protection and reforestation. This is the most common form of tenure for users of timber in United States National Forests, and is used in some form and degree by most forest management agencies in North America.

The term *cutting permit* describes minor forms of licences to cut timber that are issued for short terms and for special purposes. Authorizations of this kind are employed to clear public timber from rights-of-way and reservoirs, to salvage timber in fire-damaged forests, to supply materials for mining, to cut fuel wood, and for a variety of other minor purposes.

These common forms of property rights in timber reveal considerable diversity in their basic dimensions. However, they refer primarily to interests in timber; rights to other forest values such as livestock forage, water, recreation, and wildlife also take a variety of forms. Rights to timber and other benefits in the same forest area may be allocated in quite different forms to different parties, so that the patterns of overlapping rights may become varied and complicated.

ECONOMIC ISSUES OF TENURE SYSTEMS

Bearing in mind the multidimensional character of property rights in forests, and the wide range of property forms, we can now review briefly the main features of forest tenure systems that affect the efficiency of resource use and the distribution of benefits.

Method of Allocation

Rights to timber and forest land can be obtained in various ways. Timber-staking traditionally provided a "first-come-first-served" method of establishing private property rights in the public domain. Pre-emption and improvement was another way of obtaining grants of public lands in the frontier era; by this means early homesteaders could acquire title to the lands they used and improved. Nowadays it is common in Canada's major timber-producing provinces for corporations to secure rights over resources through bilateral negotiations with governments. In the United States, rights to harvest timber on public lands are usually allocated by competitive bidding. And, of course, rights of most kinds can be acquired by purchasing them from others.

From an economic viewpoint, the important issue is whether the method of allocation enables the resources to find their way into the hands of whoever can make the most productive use of them. Competitive allocations encourage this process, because the most efficient users can outbid the less efficient. Non-competitive allocations do not, but much depends on whether the rights are transferable once allocated. As long as they are freely transferable they will tend to be reallocated through subsequent purchase and sale to the highest users. In this case, the only long-run consequence of the original allocation method is on the distribution of income; if the subsequent transactions are competitive, all the benefits not captured in the initial allocation will be captured by the first recipient of the property rights.

Scope of Rights

The more restricted the rights held by users, the more externalities can be expected. For example, if the rights are limited to the cutting of timber, excluding rights to water, agriculture, and recreational benefits, the holder will have no economic incentive to consider the impact of his harvesting on these other values, or to take account of the external costs or benefits he imposes on them, as we noted earlier. In contrast, if his rights include the full range of values, he will internalize these costs and benefits and attempt to maximize the aggregate net return from all of them. Where the incentive to take account of some values is eliminated by the narrow scope of the users' rights, they can be protected only through governmental regulation.

Even the most comprehensive forms of property rights exclude some values. Certain externalities cannot be easily internalized, for technical reasons. Examples include the amenity of forest landscapes and values associated with fish and wildlife that migrate in and out of the relevant forest land.

Security

The security of rights to resources is a major influence on entrepreneurial investment and operational planning. Insecurity invokes risk and, as noted in Chapter 6, investors are generally averse to risk. So insecure rights impede investment in resource development and management.

A primary determinant of security is the duration of the rights. Private title carries an infinite term, while some licences and permits have terms of only a year or two. However, where the terms are finite, the holder's security depends not only on the duration of his rights but also on the provisions for renewal. If there are no rights of renewal the term is critical; if renewal is automatic the term is unimportant. Between these extremes there may be variety of provisions for qualified rights of renewal.

Against the licensee's customary desire for security lies the licensor's interest in retaining the ability to alter allocation arrangements. Landlords, both private and public, have an interest in preserving their flexibility to reallocate resources in the face of changed circumstances or to alter the terms and conditions under which rights to their resources are assigned to others.

The licensee's interest in security on the one hand, and the licensor's interest in flexibility on the other, conflict on the issue of the term of rights. A recent innovation in forest tenure arrangements neatly reconciles these interests by providing for "evergreen renewal"; that is, a provision in the licence to enable the licensee and licensor to negotiate the terms of a new licence to replace the existing one when its term has only partly expired. This technique provides a regular opportunity to change the terms and conditions while ensuring that the licensee never has to face the imminent expiry of his resource rights.

Scope for Intervention

The security of rights is also affected by the ability of governments or others to interfere with the activities of the holders and by any ambiguity about how holders may exercise their rights. The frequency of externalities in forestry gives governments wide scope for regulating forest operations in order to protect non-timber values, which is therefore an important consideration in forest tenure arrangements. Governments can employ legislation to intervene on both private and public lands, but where they are the licensors of users of public resources the terms and conditions of the licences provide them with an additional means of regulating operations.

Broad scope for regulatory control enhances the government's power to protect all social values and interests, but it restricts the freedom of property holders to use the resources in the most advantageous way for them. Susceptibility to governmental regulation thus diminishes the quality of property rights and their value. A related aspect of the quality of rights is the holders' ability to enforce them and protect them against interference from others.

Allocation of Management Responsibilities

Many forms of usufructory rights over forests not only provide their holders with rights to use resources but also assign them management responsibilities. In Canada especially, elaborate forest management agreements require the licensees to manage public forests according to plans approved by the responsible forest agencies.

It is important to distinguish the issue of who is responsible for resource management from the question of who is to pay for it. This is because licensees who assume such contractual responsibilities are often reimbursed, directly or indirectly, for the costs. The ultimate impact on the distribution of resource rents thus depends on these financing arrangements as well as more direct fiscal measures.

Distribution of Resource Rent

Taxes, royalties, stumpage charges, and other levies determine how much of the potential resource rents will accrue to the property holder. Most fiscal devices also affect incentives to conserve forests and invest in future yields. These issues are considered in the following chapter.

PRIVATE AND PUBLIC OWNERSHIP

The discussion on the preceding pages suggests that the customary distinction between private and public ownership of forests is inadequate. The property rights of private owners are almost always qualified by restrictions in their deeds and titles, by regulatory laws and regulations and by various taxes and charges. Those who utilize public forests are usually private corporations or individuals who hold some form of usufructory rights. In most jurisdictions where forestry is important, those who utilize forests do so within a tenure system that consists of a range of public and private property rights, which overlap and blend in various ways.

Public ownership of forests is common throughout western, eastern, and third world countries. National, regional, and local governments, as well as a variety of public boards and agencies, hold title to forestland. Much is managed for timber production, but special public needs, such as those associated with parks, greenbelts, and watersheds, often provide the main purpose of governmental control. In most western countries, public ownership of forests is far more prevalent than public ownership of agricultural and other types of land.

As noted early in this chapter, the widespread public ownership of forest land in the United States and Canada is the result of early political decisions to abandon the traditional procedure of granting title to land and resources sought by private users. The shift in policy toward maintaining public ownership of forest land seems to have been motivated by three economic considerations. One was a perceived threat, vigorously asserted by the powerful conservation movement around the turn of the century when western forests were opening up, that the remaining forests would fall into the hands of land barons and developers who would exploit them too fast, leaving inadequate resources to support the industrial needs of future generations.

Another concern of the U.S. conservation movement which influenced Canadian policy-makers as well was that speculators would lay claim to resources and withhold them from development in the expectation that their value would rise, to the detriment of the present generation. This fear obviously conflicts with the contemporary concerns about too rapid resource depletion, but both these propositions about undesirable rates of exploitation by private owners gave support to advocates of public ownership of land and resources.

A second motive was to protect the public financial interest in natural resources. Timber provided the foundation for early regional economies, and many people saw the expansive natural forests as a bank of public assets that could be realized fully only by retaining them for sale as the demand and value of them rose over time. It is worth noting that this view presumes that potential private purchasers have expectations about future values that are more pessimistic and less accurate than those of governments, otherwise the prices they would be prepared to pay for the resources would match the present value of the returns that governments could expect from later sales. But shortsightedness on the part of private corporations was a popular view among reformists early in this century.

The third motive, of more recent vintage, was to protect the public interest in forest values other than timber. This motive rests on the twin presumptions that private owners will pay insufficient attention to these other values and that they can be better protected through public ownership than by either creating more comprehensive private property rights or by regulating private owners.

All these arguments in favour of public ownership of forests are subjects of continuing academic and political debate. As noted in Chapter 6, economic theory does not demonstrate conclusively that the social interest lies in exploiting resources either faster or slower than would result from market forces, and experience does not reveal that governments consistently manage forests better or worse than private owners. In any event, in the United States and Canada today there appears to be a firm political commitment to maintaining at least the present degree of public ownership of forests.

This commitment to public ownership is paralleled by an equally entrenched commitment to the private sector to utilize the resources. In North America, unlike many countries that rely on state enterprises, virtually all timber, and most other forest products and services as well, are recovered and used by private corporations and individuals. These circumstances put a heavy onus on the form of property rights used to provide private users with access to public resources.

But the task of designing usufructory rights that will encourage their holders to utilize resources in the interests of society as a whole is more difficult for forests than for most other natural resources. As we noted above, the unusually long planning and investment periods in forestry and the time it takes for the full impact of activities in a forest to be felt, mean that users will have incentives to respond to all the future costs and benefits of their actions only with rights of very long duration. The prevalence of multiple uses and products, public goods, and external effects similarly complicate the design of property rights.

For all these reasons the property rights of users of both public and private forests are rarely adequate to ensure that their incentives always converge with the interests of society at large. The primary alternative to improvement in property rights systems as a means of promoting this convergence is governmental regulation of behaviour. Each of these approaches will be the most effective, relative to its cost, in responding to particular problems and circumstances.

REVIEW QUESTIONS

1 If property can be described as a "bundle of sticks," each stick representing a right, what are some of the rights that can comprise someone's property interest in a forest?
2 Explain the externality that may result if someone has a well-

defined interest in the industrial timber produced in a forest but no one has a comparable economic interest in the wildlife. Why must government take responsibility for protecting the wildlife values?

3 If someone holds only short-term harvesting rights over a forest, how are his profit-maximizing decisions likely to be less efficient from the viewpoint of society as a whole than if his rights were perpetual?
4 Why is the transferability of property important in promoting the efficient use of resources?
5 How do you explain the greater political support for public ownership of forestland than for farmland?

FURTHER READING

Coase, R.H. 1960. The problem of social cost. *Journal of Law and Economics* 3:1-44

Demsetz, H. 1967. Toward a theory of property rights. *American Economic Review* 57(2):347-59

Fortmann, Louise, and John W. Bruce (eds.). 1988. *Whose Trees?: Proprietary Dimensions of Forestry*. Boulder, CO: Westview Press. Chapter 8

Furubotn, E., and S. Pejovich (eds). 1974. *The Economics of Property Rights*. Cambridge, MA: Ballinger Publishing Co.

Libecap, Gary D., and Ronald N. Johnson. 1978. Property rights, nineteenth-century federal timber policy, and the conservation movement. *Journal of Economic History* 39(1):129-42

Pearse, Peter H. 1980. Property rights and the regulation of commercial fisheries. *Journal of Business Administration* 11(1 & 2):185-209

—. 1988. Property rights and the development of natural resource policies in Canada. *Canadian Public Policy - Analyse de Politiques* 14(3):307-20

Posner, Richard A. 1986. *Economic Analysis of Law*. 3rd ed. Boston: Little, Brown & Co. Chapter 3

Randall, Alan. 1987. *Resource Economics: An Economic Approach to Natural Resource and Environmental Policy*. 2nd ed. New York: John Wiley & Sons. Chapter 8

Scott, Anthony. 1984. *Does Government Create Real Property Rights?: Private Interests in Natural Resources*. Discussion Paper 84-26, Department of Economics, University of British Columbia. Vancouver, B.C.: Dept. of Economics, UBC

CHAPTER TEN

Forest Taxes and Other Charges

Out in the Forest . . .

On all its private forest land, Peavey Forest Products Limited pays an annual property tax. The value of the land and timber on each parcel is estimated by the tax assessor and a percentage rate is applied to that value to determine the tax that must be paid.

On behalf of his company, and with the help of its industry association, David Cameron has been lobbying the government to change this tax on the grounds that it is a serious obstacle to forestry. They have produced data showing that, even at modest tax rates, the taxes paid year after year on growing stands can easily accumulate, with interest, to amounts exceeding the value of the crops at harvesting age, wiping out any return to the forest owner. They argue that the tax also encourages owners to deplete their forests in order to reduce their liability to the tax. Moreover, owners find it difficult to pay this tax in years when they have no revenues from harvests. To overcome these problems, Cameron and the industry association propose that this tax be eliminated in favour of a yield tax on timber harvested, or a land tax based on the productive capacity of the land regardless of the timber on it.

On the public timber the company harvests under its timber sale licence it must pay a stumpage fee on each cubic metre harvested. In this case the Forest Service has proposed a change, because these charges create a strong incentive to "high-grade" forest stands and consequently put a heavy onus on utilization regulations. The Forest Service would prefer, instead, to levy the stumpage charges for the entire licensed tract as a lump sum. Cameron and other industry representatives have resisted this proposal on the grounds that it would put too much financial risk on the operator.

Governments tax private forests, and both public and private forest owners levy various charges to carve out their share of the economic returns from timber and other forest products and services when they are used by others. These fiscal devices take the form of royalties, rentals, stumpage payments, taxes, and a variety of other levies. They all influence the way forests are managed and used, and the distribution of the benefits they produce.

This chapter considers taxes and other charges that are applied specifically to forest resources. It does not deal with general income taxes, sales taxes or other direct and indirect taxes on businesses, individuals, goods, and services. Those general fiscal devices bear on forestry in the same way that they bear on other economic activity, and they are better dealt with in texts on public finance. Our concern here is with the special kinds of levies that are applied directly to forest land and timber.

ECONOMIC ISSUES IN FOREST CHARGES

Traditionally, three qualities of taxation considered to be desirable are *neutrality*, *equity*, and *simplicity*. A neutral tax is one that does not create incentives to change behaviour. It will not distort the allocation of resources or the efficiency of resource use.

Few taxes and charges are entirely neutral. Most invite those who must pay them to lighten the burden by altering their economic activity in some way. Thus forest owners, faced with levies on their property or their harvests, find it advantageous to change their decisions about the intensity of forest management and silviculture, the length of the rotation period, the degree of utilization in harvesting, and so on. It is therefore important to identify the incentives created by particular kinds of levies, the distortions in behaviour that they cause, and the economic losses or costs that result.

Equity, or the distribution of income, is important because levies on forest resources shift income and wealth from the payers to the receivers. Indeed, this is usually their primary purpose. Public and private forest owners use a variety of taxes, fees, and stumpage charges to capture economic benefits that would otherwise accrue to others. But because most charges are not neutral, they cause changes in economic behaviour which have subtle and complicated effects on the distribution of income, well beyond the obvious transfer of income from payers to receivers. As a result, the ultimate *incidence* of a levy, which refers to the final distribution of its burden, requires careful analysis.

As noted in Chapter 1, equity or fairness is a subjective concept, but it is particularly important in taxation. Two-well established principles of equity in taxation are the *ability to pay principle* and the *benefit principle*. The first of these implies that people should be taxed in accordance with their wealth and income, and those in similar circumstances should be taxed similarly. Income taxes reflect this approach. The benefit principle implies that people should be taxed according to the benefits they receive from public expenditures. Property taxes that pay for local services are consistent with this idea. These two principles are not mutually consistent, however, and modern tax systems, including taxes and charges on forests, embody a mixture of them.

Simplicity refers to the complexity of fiscal arrangements for administration and enforcement, their understandability, and costs of compliance. Some charges are very demanding of data, of econometric calculations, or of resources required to administer and police them. These costs are often significant and have efficiency and distributional consequences of their own.

Another important consequence of forest levies is their effect on stability of incomes, both of the payer and of the receiver. Some levies, such as stumpage charges on timber harvested, increase and decrease with the income generated by the payer. Others, like land rentals and taxes, tend to be more constant over time, regardless of payer's level of production and earnings. The latter provide more stable income to the receiver, but the former bear a more stable relationship to the payer's earnings and ability to pay. Depending upon the way a charge is levied, instability and risk can be shifted between payers and receivers.

Thus taxes and other levies have important implications for efficiency in the way resources are allocated and used, the distribution of income, administrative complexity, and stability of incomes. This chapter focuses on these effects of forest taxes and charges.

TYPES OF FOREST LEVIES

Levies on forests take various forms and each form has particular economic impacts. The most common types are classified in Table 5 according to their tax or assessment base. As shown in the left-hand column of the table, the most common bases are the land, the timber, the harvest, and the property right. The last of these refers to charges for the acquisition or renewal of property rights over the forest, such as fees for licences and leases.

Land Taxes and Rentals

Governments commonly tax private forest land, and both public and private landowners often require rental payments from those who

TABLE 5: Common forms of levies on forest resources

Base	Common form	Usual method of determination	Usual method of assessment
Land	land rent	arbitrary	annual levy
	land tax	percentage rate on land value or productivity	annual levy
Timber	property tax	percentage rate on timber value	annual levy
Harvest	royalty, severance tax, and cutting fees	specific fees for volumes harvested	according to harvest
	yield tax	percentage rate on value harvested	according to harvest
	stumpage	competitive bidding or appraisal	lump sum, annual charge or according to harvest
Rights	licence fee	more-or-less arbitrary	lump sum or annual charge

occupy or use their forest. Rentals are commonly paid by holders of forest leases and licences to public or private landlords, and land taxes are usually levied by governments on freehold owners of land. Both land taxes and rentals invariably consist of fixed amounts that must be paid annually, so they have similar economic effects.

These levies are usually determined in one of three ways. Rentals are typically a more-or-less arbitrarily specified amounts payable per hectare of licensed land. Land taxes are often assessed by applying a percentage tax rate against the estimated value of the land. In other cases, both taxes and rentals are determined by applying a rate against the value of the land's productivity.

The last two of these methods of assessment converge as long as the land is put to its most productive use. That is, the market value of the land and the economic productivity of the land should both equal the site value described in Chapter 7.

Forest land taxes based on the sustainable yield or some other measure of land productivity are often levied on private lands dedicated to continuous forest production. For example, in some jurisdictions of North America and Scandinavia the tax base is the estimated mean annual increment of the land multiplied by current timber prices. Under this formula, even if the tax rate remains constant, the taxes payable fluctuate as timber prices change over time.

Land taxes and rentals have the rare quality of neutrality. Because the amount payable does not depend on the inventory, the amount harvested, the choice of rotation, or other variables that can be manipulated, it does not create incentives to alter management decisions and so does not distort efficient resource use. Whatever management regime maximizes private returns without the tax or rental will continue to maximize returns to the owner with it; the levy is neutral in this sense. Its effect is only distributional, shifting resource rent from forest owners to governments or from tenants to landlords.

The tax rate used in assessing a tax on land value or productivity can be adjusted to capture any desired proportion of the economic rent, and if the tax is consistently administered it will capture a consistent share of the rent across forest lands of varying productivity and value. In contrast, rental fees, particularly those levied by governments on holders of usufructory rights in public forests, are usually assessed at uniform rates, and so capture an inconsistent proportion of resource rents.

A land tax or rental is neutral, as noted, as long as it captures less than the entire economic rent, but if it exceeds the rent it will make the land unprofitable to use at all. Thus a uniform levy on forest land of varying productivity will shrink the extensive margin of forestry. In terms of the discussion in Chapter 5 and Figure 15, a levy on marginal land will impose a loss on the owner, driving the land out of productive forestry use and, if it is not levied on land used for other purposes as well, into other uses.

Levies on land and land values tend to be fairly stable because they do not depend on the level of production or earnings, though some forms vary with timber prices as noted. Finally, such taxes and rentals are typically simple to calculate, administer and collect. Data requirements are demanding only in the more elaborate attempts to estimate land productivity.

Taxes on Timber

Property taxes are often levied on standing timber. Typically, a percentage tax rate, or mill rate, is applied against the appraised value of the timber inventory. Appraisals are usually based on information about recent sales of comparable timber or calculations of recoverable volumes and values.

Property taxes of this type generate assessments that are proportional to the current value of the forest crop. Other things being equal, they increase as the forest grows in volume and value. The forest owner is therefore faced with repeated annual assessments on his timber, increasing in amount, for as long as he continues to grow the crop. This situation discourages owners from producing valuable forest crops, and such taxes are therefore often considered detrimental to forest management.

Even modest tax rates applied to growing forests can generate payments that, with cumulating interest, will capture a substantial portion of the value of the crop by the time it reaches harvesting age. For example, if the stand described in Table 2 of Chapter 7 were subject to a tax of only 1 per cent per year, the tax payments and interest on them would accumulate to nearly one-quarter of the value of the crop by the time it reached the harvesting age of fifty-eight years. With less productive land, or a higher tax rate, the tax could wipe out any gain to the owner and may impose a loss. As a result, forest owners sometimes find it advantageous to keep their land denuded of valuable forest. By thus reducing the profitability of forestry, and by lowering the returns to growing timber relative to the returns to other land uses, such taxes may drive land out of timber production altogether.

Governments have sometimes sought to take advantage of the incentive to harvest resulting from a property tax on timber. Especially in developing countries and regions with extensive private holdings of virgin forests, such taxes have been used to encourage harvesting and expansion of the forest industry. However, if the tax continues to apply to subsequent managed crops it will tend to discourage continuing timber production.

Aggravating these effects, property taxes fail to synchronize tax liabilities with the owner's revenues from his crops. In contrast to levies on harvests, these taxes recur every year the crop is grown, imposing on owners the added burden of financing tax payments.

Taxes of this kind are usually easy to administer because they correspond to property taxes applied to other real estate and are usually integrated with general property tax administrative arrangements. They call for data relating to current timber inventories and values, both of which must be revised with each assessment. Because the assessments depend on current prices and volumes, they vary with the value of the owner's assets, but not necessarily his income.

Property taxes on forests normally apply to both land and timber and both are assessed together, though not always according to the same criteria or rates. The economic effects are a combination of those associated with property taxes on land and on timber, outlined above.

Levies on Harvests

A considerable variety of taxes and charges are levied on timber harvested, but we can classify them into three forms according to the way they are assessed, which governs their economic effects. One category consists of assessments specified in dollars payable on the *quantity* of timber harvested. This includes royalties, severance taxes, and cutting fees or cutting dues. The second takes the form of a percentage of the *value* of timber harvested, usually called a yield tax. The third is a charge in the form of a share of the *value* of the standing timber to be harvested, referred to as a stumpage price.

The terminology used for various types of taxes and charges is often confusing and inconsistent. In the following paragraphs each is defined in a specific way in order to facilitate discussion of its economic effects. However, note that these definitions are not universally adopted.

A basic distinction should be made between a tax on the one hand and a royalty or stumpage price on the other. A tax is always a government levy which, in effect, expropriates for the public purse a portion of private wealth or income; a royalty or stumpage price is a purchaser's payment for a commodity—timber. However, as we shall see, the incidence or impact of taxes and other levies may be similar, depending upon how they are assessed.

The three general types of charges on timber harvested can be described as follows:

(a) *royalties, severance taxes, and cutting fees or dues*. These are typically specific dollar charges, or schedules of charges for various categories of timber, levied on each cubic metre or other unit of timber harvested. In most cases they are modest charges, and being specified in money terms they tend to be eroded by inflation over time.

Traditionally, a royalty represented the value of a resource that the sovereign reserved to himself when granting rights of ownership or use to a private person, as described in the preceding chapter. Severance taxes, most common in the United States, were conceived as payments to society for depleting a natural resource. Cutting fees and dues, like royalties, are usually simply schedules of charges for public timber. All these levies are fixed in dollars per unit of timber harvested and so, regardless of their original intent or justification, they have similar economic effects.

The direct distributional effect of these levies is to shift the economic returns from timber from the harvesters to the government. They also affect economic efficiency and the allocation of resources in harvesting by making it unprofitable for the owner to harvest marginal stands of timber and to recover marginal logs.

The value, per cubic metre, of logs recoverable from different stands, different species, different trees and even from individual trees varies widely. Figure 24 illustrates how the value of logs, per cubic metre, typically declines with their quality or grade. At some grade, g in Figure 24, the log value is just sufficient to cover the harvesting cost, indicating the marginal or "cut-off" grade of log, and the limit of profitable harvesting. A royalty or tax of a fixed amount per cubic metre harvested has the effect of lowering the net value of all logs to the owner by the amount of the assessment, reducing the range of quality that can be profitably recovered, by g_tg in Figure 24. Such levies thus encourage "high grading" in timber harvesting.

This contraction of the intensive margin of recovery is paralleled

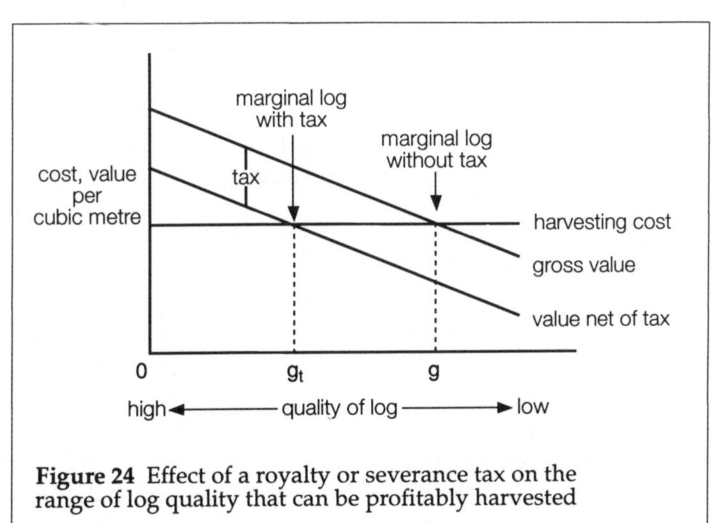

Figure 24 Effect of a royalty or severance tax on the range of log quality that can be profitably harvested

by a contraction of the extensive margin as well; that is, stands of timber at the margin of economic recoverability are made sub-marginal by such levies. This development can be illustrated with reference to Figure 8 in Chapter 3, where a levy on all timber harvested would have the effect of lowering the value of the inventory and shrinking the extensive margin. Correspondingly, the growing of forests on marginal sites is also rendered unprofitable. This effect can be illustrated with reference to Figure 13 in Chapter 5, where a levy on timber will lower the marginal revenue product of labour on any site, thus constraining the intensive margin of forestry. In all these ways, such levies impede efficient forest management and use.

Royalties and other charges of this kind are relatively simple to administer because they require information only about the timber removed from the forest, which is usually collected for other purposes as well. The payer's liability to these assessments depends on his harvests, which facilitates his ability to pay, though at the expense of stability of revenues to the recipient.

(b) *yield taxes*. Yield taxes are government levies on timber cut on private land, taking the form of a percentage rate applied to either the selling price of the logs (the *gross value* of the harvest) or the stumpage value of the standing timber (the *net value* of the timber harvested). They are thus *ad valorem* taxes, being based on the value of the tax base—the harvest. Yield taxes are not widely used in North America, but they are sometimes adopted as an alternative to property taxes on timber.

For the forest owner, a yield tax levied on the sale value of logs has the same effect as a lower selling price. It constrains the intensive margin of profitable harvesting by making marginal logs unprofitable to recover. The effect is similar to that of a royalty or severance tax, illustrated in Figure 24, except that the yield tax reduces the gross value of logs by a constant proportion rather than a fixed dollar amount.

This distortion of the economic margin does not occur when the tax is levied as a percentage rate applied to the *net value* of the timber. At the margin, the value of timber is zero and a tax assessed on that value is zero also, so the margin does not change. In this respect the economic effects of the tax are identical to those of a stumpage levy, described below.

Yield taxes are more complicated to administer because they call for information not only about the quantity of timber harvested but also about its market value. Procedures for collecting and averaging price data, allowing for fluctuations and policing sale information, add considerably to the cost of assessment and collection.

As an alternative to annual property taxes on timber, yield taxes have the advantage of avoiding the accumulation of payments and interest on payments as the forest grows, with the resulting financial disincentives to carrying crops noted earlier. They enable governments to take a consistent share of the value of forest production when it is produced, facilitating the payer's ability to pay. However, both the level of harvesting and the price of harvested products tend to fluctuate in response to market cycles, so the revenue from yield taxes tends to be highly unstable. For local governments which usually depend heavily on revenues from taxes on property such instability presents difficulties because they have continuous financial requirements. This explains the more widespread use of property taxes on timber, which produce more stable revenues.

(c) *stumpage charges*. Stumpage prices or charges are payments made to a public or private forest owner for timber harvested by someone else. In this respect they serve the same purpose as royalties and severance taxes, but the latter are usually relatively modest fixed rates applied uniformly, while stumpage charges are typically more discriminating levies that are intended to capture for the owner all or a substantial share of the value of timber which differs on each tract harvested.

Stumpage charges are based on the *net value* of the timber as it stands "on the stump." They can be designed to capture all or a portion of this net value.

Stumpage prices are determined in a variety of ways. In the United States most public timber is sold at competitive auctions, where potential buyers are invited to bid for the timber in terms of dollars per unit of volume harvested. Bids are submitted either orally or by means of sealed tenders, usually subject to a minimum acceptable price or "upset price." The highest bidder wins the harvesting licence and thereby sets the stumpage charges. Competitive bidding is used to allocate certain minor forms of cutting rights in public timber in Canada as well, and it is common practice for private forest owners to use similar procedures in selling timber. As long as competition is sufficiently vigorous, this method provides the seller maximum assurance that he will collect the full net value of each tract allocated for harvesting.

In the absence of competition among buyers the upset price becomes the stumpage price, and the share of the net value of the timber captured by the seller depends heavily on the way these administered prices are calculated. Appraisal procedures vary, but they have certain common elements. First, the volume of each species and grade of logs recoverable from the forest tract is estimated from inventory data. Second, the value or selling price per cubic metre of each category of logs is determined from market information or from calculations of the value of the products that can be manufactured from them. Third, the cost of harvesting the timber per cubic metre, including costs of road-building, logging, and transporting the logs to the market where the selling prices can be realized, are estimated and subtracted from the selling price. Finally, the remainder, referred to as the "conversion return," is reduced by an allowance for the operator's profit and risk to yield the stumpage price payable to the seller of the timber.

These calculations can be illustrated as follows:

	value or cost per cubic metre
selling price of the logs	$100.00
minus cost of logging and delivery	-60.00
equals conversion return	40.00
minus allowance for operator's profit and risk	-15.00
equals appraised upset price	25.00

The calculation is either made separately for each species of timber, or an average is calculated by weighting the selling price of each species by its percentage of the total volume of the timber to be harvested.

Appraisals of this kind are exceedingly complex and highly demanding of data relating to the timber and operating conditions, current costs, and market prices. They also depend on judgments about such matters as allowances for risks, depreciation, and reasonable returns to operators. As a result, the administrative costs tend to be high and the results are often contentious.

The greater the proportion of the net value of timber the stumpage charges are intended to collect, the heavier the onus on these calculations. In some cases owners use relatively simple techniques to set stumpage rates that will secure for them a reasonable share of the resource values, such as by extrapolating values from competitive sales of comparable timber in the region or by rules of thumb for dividing revenue from the sale of the logs between the owner and the operator.

The economic efficiency effects of stumpage charges depend not only upon how they are determined, but more importantly upon how they are assessed. Once the stumpage price for a tract of timber is determined, it may be applied to the estimated volumes of standing timber and assessed as a lump sum, or the same amount may be collected in equal annual payments over the planned harvesting period, or the stumpage rate may be assessed on the timber harvested as it is cut and scaled, or it may be collected in a variety of other ways. If the charges are payable in a lump sum, in predetermined annual amounts, or in any other way that is unaffected by the entrepreneur's actual harvesting, his economic behaviour will not be influenced by the charges regardless of how they were determined in the first place. They become neutral fixed or sunk costs, affecting none of his production decisions at the margin. His profit-maximizing pattern of production will be the same with them as without them.

In contrast, stumpage charges assessed uniformly on the timber actually recovered from the forest, in dollars per cubic metre of logs removed, will affect the operator's decisions. In this case his net return on each cubic metre is reduced by the amount of the stumpage levy, rendering marginal timber unprofitable to recover through the same effect as a royalty or severance tax, illustrated in Figure 24. However, because stumpage charges are usually higher, the effect is correspondingly greater.

This incentive to "high-grade" timber in logging operations can have significant consequences when stumpage prices levied on the logs recovered are aimed at capturing all or most of the net value of the standing timber. If the stumpage price per cubic metre is set at the average net value of the timber in the stand, the operator will earn a profit only on timber worth more than the average. On the other half of the economically recoverable volume he would earn a loss, and so his financial incentive is to refrain from harvesting it. Some regulatory agencies impose utilization standards in order to offset these incentives. But such rules are inevitably more or less arbitrary, and rarely take account of the variations in the marginal log in different stands and logging conditions, or with changing costs and prices.

In short, stumpage charges levied on the *logs recovered* from the forest, in contrast to assessments on the standing *timber allocated* for harvesting, constrain the intensive and extensive margins of profitable operations, resulting in waste of potentially valuable timber.

Assessments on the standing timber are thus preferable on grounds of neutrality. But assessments on the volume removed call for less exacting data about the forest itself, and remove some risk from operators. Moreover, logs recovered from the forest are usually scaled for other purposes in any event, so it is often convenient to base stumpage charges on these measurements.

Stumpage charges fixed in advance of harvesting leave the payer vulnerable to any miscalculations or mistakes in determining bids or administered prices. They also impose on him all the risks associated with unforeseen changes in harvesting conditions, costs, and timber prices. These risks are greater the longer the period between the commitment to make the payments and the completion of harvesting. They are most onerous when the charges are paid as a lump sum in advance, and the harvests extend over a long period.

The burden of risk borne by the operator can be alleviated in various ways. One is by designing the assessments to capture only a portion of the net value of the timber harvested, so that the risks will be shared proportionally between the owner and the producer. Another, already noted, is by basing the payments on the timber actually recovered rather than fixing them in total in advance, thus removing the financial uncertainty about recoverable volumes. And the risk of changing market conditions can be offset by tying the assessments to some index of product prices. Thus some public agencies adjust stumpage charges, determined at the beginning of a harvesting contract, according to a sliding scale as the market price of specified forest products changes. All these measures shift risk from payers to receivers of stumpage payments.

Charges for Rights

For completeness, we must consider charges for rights to timber, as distinct from charges for the timber itself. It is usual for governments to levy a fee for forest licences and leases which are independent of the timber actually cut by the licensee or lessee. These charges are usually modest, and are typically justified as payments by licensees to have resources reserved for their exclusive use (or, conversely, as payments to the government for withholding the resource from use by others).

Such fees are often payable when licences are issued, or as fixed annual assessments to maintain the rights in good standing. Initial once-and-for-all licence fees have the same neutral effect on behaviour as any other lump sum levy, and annual fees are neutral in the same way as annual rentals, described earlier. As long as the payments are not affected by the licensee's behaviour, they will not distort efficient operations.

COST OF BEARING RISK

Throughout this discussion of taxes and charges, we have referred to the way risk, and the burden of changing economic circumstances, is borne by forest owners and users. Clearly, the way this burden is shared is affected by the way taxes and other charges against timber are levied. Assessments that fix the amount and timing of charges in advance minimize the payer's uncertainty about his revenues, but they impose on him all the risks of miscalculation or unforeseen events. Conversely, to the extent that payments are based on the timber that the operator actually recovers, and the market conditions he encounters, his risks are reduced and those of the receiver correspondingly increased. As we have seen, various techniques can be employed to distribute risk between payers and receivers.

The importance of this matter derives from the general rule that entrepreneurs are averse to risk, as discussed in Chapter 6. As a result, they demand higher expected returns from risky ventures than from secure ones. It follows that if those who use the forest are forced by fiscal means to bear more risk and uncertainty, the rents that can accrue to the forest owner will be lower. In the often unstable economic circumstances of the forest industry, this effect can be significant.

If a private entrepreneur has a choice between a tax fixed once-and-for-all and another that is expected to yield the same amount but is based on his profits, he is likely to prefer the latter. It is less of a threat in the event of unexpected adversity and low returns because it shares the risk with the government and so is less burdensome. For this reason, it is often argued that governments can collect higher revenues from public forests in the long run if they design their taxes and other charges to shift the financial risk and uncertainty from enterprises that must pay them onto governments themselves.

In this connection, it is also important to note that governments are usually considered to be less vulnerable to fluctuations in any single source of revenue, such as a tax on forests, than are enterprises that must pay it. Governments have an interest in stable revenues, but forest revenues are often a minor component in a government's large and diverse revenue system, so that fluctuations in them alone are not highly disruptive. In contrast, a firm in the business of harvesting timber is likely to be much more sensitive to changes in such charges. Thus the cost to a government of assuming the financial risks of a forestry operation is likely to be less than the gain to a forest enterprise from avoiding them.

The susceptibility of an enterprise or government to the instability of payments on a particular forest tract is therefore dependent on its diversity and scale of operations. The same is true of the burden of financing charges, such as annual taxes on the crop, until harvesting takes place; if the crop is part of a large forest in which crops are continuously harvested the financing problem is alleviated.

OTHER ECONOMIC CONSIDERATIONS

As we have seen, the only neutral levies on forest resources are those that capture economic rent, such as taxes on bare land values, annual rentals, and lump-sum assessments, which are not affected by management decisions. All others generate incentives to alter production decisions in order to reduce the burden of the charges. We have noted how charges based on the volume or gross value of timber harvested narrow the range of material that timber harvesters find profitable to recover. In Chapter 7, we observed how taxes on the forest inventory give owners incentives to shorten forest rotations while taxes on harvests create incentives to lengthen them. And any charges that reduce the economic return from growing timber constrain the extent of profitable investment in silviculture and management. These are all impacts on the intensive margins of forestry.

Most taxes and charges affect the extensive margins of forestry as well. Levies on timber harvested make marginal stands uneconomic to harvest and taxes on forest inventories can make it unprofitable to grow timber on marginal sites.

The concept of neutrality therefore applies as well to the allocation of land among uses, discussed in Chapter 5. Only a property tax that imposes the same burden on land regardless of its use can avoid distorting patterns of land use. Such a tax cannot, of course, be based on timber or any improvements to the land.

On the other hand, governments often want to influence land allocation and use taxes for this purpose. Forests, like farms, are sometimes taxed at relatively low rates to encourage rural land use, stimulate employment, improve the welfare of low-income groups, or stabilize rural communities.

A related concern that has historically influenced the design of taxes and charges on timberland is about speculative holdings of timber rights. The history of the forest industry in North America is rich in examples of investors seeking to acquire, and hold, extensive rights over timber in anticipation of growing scarcity and increasing resource values. Governments have tended to view this practice with disfavour and have sought to discourage it. As noted earlier, property taxes have sometimes been used to encourage owners to harvest timber. For the same reason annual rentals have been levied on licences and leases over public timber to make it unattractive to hold rights to resources in excess of the quantities needed for planned production. This policy calls for levies on forests that extract more than the present worth of anticipated gains that speculators can expect from holding timber.

Economics offers no support for the presumption that speculative acquisition and holding of resource rights is contrary to the public interest. Speculators usually can be assumed to assist in the efficient allocation of resources over time by constantly trying to identify the most advantageous time to utilize them. Nevertheless, if it is considered desirable to discourage speculative activity, levies on timber rights afford an expedient means of doing so.

Finally, it must be emphasized that the impact of any particular tax or charge on forest resources is affected by the whole fiscal and institutional environment within which forest production takes place. Some levies are allowed as deductions in calculating income taxes or capital gains taxes, others fill loopholes or offset incentives that would otherwise distort activity, still others compensate for inequities that would otherwise exist, and so on. These circumstances vary, and so the effects of forest levies must be analysed with careful attention to the circumstances of each time and place.

REVIEW QUESTIONS

1 What is a "neutral" tax? Give an example.
2 Using the data presented in Review Question 3 of Chapter 7, calculate the amount to which tax payments and interest on them would accumulate if a tax of 1.5 per cent were levied on the value of the timber every year until the stand reached the age of fifty years. (For simplicity of calculation, assume that the timber volume indicated for each fifth year remains constant at that level for the following four years.)
3 How does a yield tax affect incentives to invest in silviculture?
4 In what way will the extensive margin between forestry and agriculture be affected by a land tax that applies at a higher rate on forestland than on farmland?
5 How will a logger's incentive to "high grade" a stand be affected by stumpage charges that are (a) assessed on each cubic metre of timber recovered, and (b) assessed as a lump sum for the stand regardless of his actual harvest?

FURTHER READING

Boyd, R.G. 1986. *Forest Taxation: Current Issues and Future Research*. Report to the Southeastern Forest Experiment Station Research Triangle Park, NC. USDA Forest Service

British Columbia Royal Commission on Forest Resources. 1976. *Timber Rights and Forest Policy in British Columbia*. Report of the Royal Commission on Forest Resources, Peter H. Pearse, Commissioner. Victoria: Queen's Printer. Chapter 13

British Columbia Task Force on Crown Timber Disposal. 1974. *Timber Appraisal: Policies and Procedures for Evaluating Crown Timber in British Columbia*. Second Report of the Task Force on Crown Disposal, July 1974. Victoria: B.C. Forest Service

Boyd, Roy G., and William F. Hyde. 1989. *Forestry Sector Intervention: The Impacts of Public Regulation on Social Welfare.* Ames: Iowa State University Press. Chapter 7

Duerr, William A. 1960. *Fundamentals of Forestry Economics*. New York: McGraw-Hill. Chapters 26 and 27

Forest Products Research Society. 1988. *Forest Taxation: Adapting In An Era of Change*. Proceedings 47352: Papers presented at a conference in Atlanta, Georgia, 20-2 May, 1987, Margaret P. Hamel, ed. Madison, WI: Forest Products Research Society

Gregory, G. Robinson. 1987. *Resource Economics for Foresters*. New York: John Wiley & Sons. Chapter 8

Musgrave, Richard A., and Peggy B. Musgrave. 1984. *Public Finance in Theory and Practice.* New York: McGraw-Hill. Part 3

CHAPTER ELEVEN

Developments in Forestry Economics

Out in the Forest . . .

For the previous generation of managers at Peavey Forest Products Limited forest planning and economic analysis were simple. They depended on their knowledge and experience to make judgments about the best harvest rates, the timber supply outlook, and the benefits and costs of forestry programs. However, the present generation discovered tools for more exacting methods. By adapting computer-based models of stands and whole forests they are now able to analyse systematically and routinely the forestry consequences of the alternative forest management decisions they might take.

Recently, Ian Olson has begun to supplement his forest planning models with economic data and programs for analysing the economic implications of alternative courses of action. He never relies on the precision of the results of these analyses, because he knows that the basic data, the cause-and-effect relationships, and the assumptions built into the models are often weak. Nevertheless, the new modelling systems enable him to take account of much more economic information, and to compare alternative forest management choices much more thoroughly and consistently than was previously possible.

This chapter pulls together several threads from earlier chapters to put them into a broader analytical perspective. It thus provides an opportunity to re-emphasize certain fundamental concepts in forestry economics. At the same time, some of the new directions of forestry economics and the developing techniques of analysis are noted.

FRAMING ECONOMIC QUESTIONS

In the first chapter of this book we referred to the hierarchy of objectives that bear on the way forests are managed and used, from the highest and most general goals of public policy to the detailed field objectives of forest workers. Corresponding to this hierarchy is one of decision-makers. Those who make decisions about the general objectives of economic policy are usually national or regional governments, depending on their constitutional responsibilities. The interpretation of such general social goals into a complementary set of objectives for the development and use of forest resources is typically the task of legislatures and cabinets. Within this framework, operational objectives for managing specific forests are set by corporations and public agencies having responsibility for the resources. Objectives for field operations are ordinarily the responsibility of supervisory personnel.

A third hierarchy consists of the instruments or means used by the decision-makers at each level to express their objectives and implement them. Thus, paralleling the hierarchies of decisions and those who make them is a structure of decision-making techniques, ranging from national constitutions down through legislation, regulations, policies of corporate boards of directors, directives, administrative rules and procedures, and on-site field decisions.

One implication of this third hierarchy is that every economic problem to be analysed must be carefully cast in its appropriate context, with reference to the scale of the decision to be made, the scope of its impacts, and the interest of the decision-maker. Decisions about general governmental policy can be expected to affect a whole jurisdiction, while others affect only a region, a forest, or a stand. These are differences in spatial scale. Decisions also vary in their temporal scale, that is, the scale of time over which actions and outcomes work themselves out. For example, the effects of a decision about reforestation on timber supply and economic returns are usually felt only after many decades, while those resulting from a decision about how a stand is to be harvested are much more immediate.

Moreover, in forestry it is particularly important to recognize the interdependence of decisions taken at different scales. For example, decisions about long-term timber supply objectives are usually made for regions, or large forest management units, but they can be met only through decisions made about how individual stands are managed and harvested. Correspondingly, short-term management decisions affect long-term decisions and vice versa. To embrace all the relevant impacts of a management decision, an economic analysis must be framed with the appropriate spatial and temporal scales.

Each decision-maker represents, or acts on behalf of, a person or body of people sometimes referred to as the *referent group*. Thus the referent group of a government is normally taken to be the people within its jurisdiction; a corporation represents the interests of its shareholders; and an individual landowner can be expected to respond to his particular interest.

If the spatial or temporal scale of the decision-maker's interest is too restricted to encompass the full impact of his actions, the result is externalities of the kind noted in Chapter 2. Numerous examples of such market failures have been noted in preceding chapters. A firm with only short-term rights to harvest a tract of forest lacks incentives to consider the long-term effects of its actions on future crops; the forest protection efforts of one forest owner may benefit neighbouring landowners; a corporation single-mindedly managing a forest for timber production may incidentally affect, beneficially or adversely, the interests of recreationists; and so on. Analyses of economic decisions from a broad social perspective must take account of such externalities, which often causes social evaluations to diverge from those of particular private interests. So it is important that the referent group, whose interests are represented in the accounting of costs and benefits, is clearly identified in each case.

Later in this chapter we discuss how to develop techniques for investigating forest management problems at different levels of analysis. But first we must refer again to the nature of the economic analysis involved in addressing forest management problems.

MARGINAL ANALYSIS

Early in this book we noted that most economic problems in forestry involved decision-making at margins, focusing attention on marginal adjustments of inputs and outputs in forest management. The issues considered in subsequent chapters drew attention to several distinct margins of adjustment.

The most familiar types of marginal adjustments in economics are those that govern the use of inputs in production processes. As shown in Chapter 2, efficient production of a given level of output calls for substituting one factor of production for another to find the combination at which their marginal rate of substitution is just equal to the ratio of their costs. And the efficient level of output is found by balancing the marginal cost of inputs with the value of their marginal product.

These rules determine the intensity of efficient forest management, that is, the relative amounts of land, labour, and capital that must be combined to produce forest products and services most efficiently. In Chapter 5 we saw how efficient factor proportions dictates "intensive" management of forest land, meaning relatively large proportions of labour and other factors relative to land in forest production, on land that yields high returns to other factors, and less intensive land use where the returns to other factors is lower.

We considered, as well, the efficient combination of outputs to be produced from a tract of forest. This led to the rule that the marginal rate of transformation, one product for another, must just equal the ratio of their values at the margin. In this case the margin refers to the trade-off among products rather than factors of production.

The question of efficient allocation of land among uses raises the issue of the "extensive" margins of forestry, considered in Chapters 5 and 8. This is the economic frontier beyond which forestry cannot generate a positive return, or generates a lower return than other uses of the land. Indeed, we noted the possibility of there being two such extensive margins: one delineating the limit of economic recoverability of existing timber in a developing region, the other indicating the margin within which forestry in the sense of growing and cultivating forests is economically advantageous.

Other chapters have dealt with the important time dimension of forestry decision-making, and the temporal margin of adjustment. Questions such as how long to grow forest crops, and how to spread harvests over time, examined in Chapters 7 and 8, involve balancing the growth in costs from one year to the next against the increment in values being generated.

Forest managers often deal with problems that bear on several of these margins simultaneously. For example, a silvicultural plan for a forest stand raises questions about what treatments to carry out, what combinations of labour and equipment to use in each task, when the treatment is most advantageous, and how much of it is warranted. These are all questions of degree, and whether the manager resolves them through sophisticated empirical analysis or by depending on his judgment and mental capabilities in the field, they call for marginal analysis of the costs and benefits involved.

Economic analysis thus focuses on the marginal adjustments in forestry. And marginal analysis draws attention, in turn, to a major practical difficulty in analysing economic questions of forest management, namely the large number of alternative courses of action, the degree and timing of each, and the variety of possible combinations of them, each of which has unique economic implications. Economic analysts can play a useful role in drawing attention to the range of possible choices faced by forest managers and policy-makers and by demonstrating the implications of them. But this often makes heavy demands on biological and economic data, on mensurational techniques, and on computational capabilities.

For this reason forest economics has traditionally been concerned with relatively simple and straightforward forest management problems, such as evaluating single, discrete silvicultural measures applied to a forest stand or determining the most advantageous age to harvest. However, the advent of widely accessible modern computer technology during the last couple of decades has expanded enormously the range and complexity of forest management problems that can be analysed. Sophisticated computer-based analytical models have now been developed in considerable variety, and many are sufficiently flexible and inexpensive to provide forest planners and managers with practical decision-support systems.

The design of complex analytical systems is beyond the scope of this book, but advanced forest models can incorporate the economic principles and evaluation techniques described in preceding chapters and can be usefully employed to reveal the economic as well as the biological and other implications of forest management plans. The rapid development of computer technology has converged with new pressures on forest resources, increasing in both intensity and variety in recent years. These new pressures and opportunities have complicated decisions about how resources are to be managed and used and have put heavier onus on economic evaluation of the alternatives. As a result, economic analysis is rapidly becoming a central and routine ingredient in forest management and is considerably easier to carry out in a practical and useful fashion.

LEVELS OF ANALYSIS

Earlier in this chapter we alluded to the varying spatial and temporal scales of analysis needed to investigate economic problems in forestry. In light of the rapid developments in analytical systems it is helpful to distinguish among these levels of analysis and note the particular kinds of question for which each is best suited.

Five major levels of economic analysis are relevant in forestry: the stand, the forest, the region, the jurisdiction, and the economy. These categories are intimately linked and interdependent, but they call for differing scales of analysis and different types of analytical systems.

Stand analyses. The forest stand is the basic unit for operational planning in forestry. Each stand presents particular characteristics, problems, and opportunities for forest managers, and it is at this level that site-specific planning is done and most operational problems of harvesting and silviculture are dealt with. The planning problem is typically to identify the optimum management regime for the stand with reference to certain objectives. Thus the problem might be to specify the sequence of silvicultural treatments that will maximize net returns.

Management decisions at the stand level traditionally have received most attention from forest economists. The principles of evaluation, described in Chapter 6, and the techniques of benefit-cost analysis suitable for this purpose are well recognized. Countless empirical economic analyses of stand-level management projects and plans have been carried out, but the results usually apply to a specific site and cannot be easily transferred to other stands that have different characteristics and are managed for different purposes.

However, computer-based analytical models now being developed enable forest managers to specify the characteristics of particular stands and flexibly and quickly analyse the economic implications of alternative courses of action according to a variety of economic or other criteria. Some of these models, referred to as *optimization models*, are capable of searching for and defining the management regime that will best serve a specified objective. Given an objective such as maximum net present value of timber production, these models are capable of defining all the marginal adjustments that are needed to serve the objective most efficiently.

Forest analyses. Management planning takes place as well at the level of the forest or management unit, which consists of many stands. Problems such as harvest scheduling, development and protection programs, and some forestry planning are addressed at this level.

Forest level analyses call for models of the kind referred to in Chapter 8, which integrate the dynamic variables of forest growth and depletion, and the influences of silviculture and protection efforts on forest inventories and yields. This level of analysis is not independent of decisions at the stand level, of course; rather, it involves assessment of the impact of activities in all the component stands with reference to objectives for the forest as a whole. Planning at each of these two levels is constrained by planning at the other.

Forest level planning models have received a great deal of attention, particularly for purposes of harvest scheduling in public forests and regulating private forests. Only recently, however, have many of these systems incorporated economic analysis.

Many of these forest planning models are now capable of incorporating large quantities of information not only about the forest itself but also about the economic conditions of production, enabling planners to explore alternative management strategies in considerable detail. Some of them are based on the mathematical optimization technique of linear programming, and have acronyms like TIMBERAM, MUSYC, and FORPLAN. These optimization models have some basic features in common. The *objective function* is the variable that is to be maximized. For example, the objective function might be to maximize the present value of the forest. The *exogenous variables* in these models refer to data that are determined outside of the model, such as interest rates, treatment costs, and product values. *Endogenous variables* are calculated by the model itself, such as growth and harvests. A variety of *activities* describe the possible treatments that can be applied to individual stands, such as planting, thinning, and harvesting. A sequence of activities applied to a stand, at specific dates, comprises a *program*. With these models, alternative strategies for managing the forest can be simulated, and their contribution to the objective function can be examined and compared.

Often the decision-maker's objectives are not fully described by a single objective function; other, secondary objectives are then formulated in terms of *constraints*. Sometimes an "even-flow" constraint is applied, restricting the range of solutions to those that generate a yield within specified bounds or within a certain percentage of the harvest in the previous period. A multiple-use constraint may require that a certain minimum area of forest of a particular type must exist at all times. Or, limits might be set on the distribution of age classes at the end of the planning period to provide for continuing yields.

Even with constraints, the number of technically possible combinations of activities and programs that can be adopted for a forest is almost infinite. For practical purposes they can usually be reduced to a manageable number for a high-speed computer using algorithms designed for such problems, and the alternative that best serves the specified objectives can be identified.

A valuable attribute of linear programming models is that they can reveal the opportunity cost of such constraints. This has enabled a major advance in economic analysis of harvest scheduling, among other issues. For example, if a minimum harvest constraint is imposed the model can calculate its "shadow price," or the extent to which it reduces the objective function. This information enables much more informed decisions about whether the harvest regulation generates benefits that justify the cost, and whether one harvest schedule serves an objective more efficiently than another.

Analyses at the level of the stand and the forest are the main interests of forest managers and forest economists. However, a broader scale of analysis is needed for certain purposes.

Regional analysis. Some economic problems are addressed at the level of the economic region, notably employment, the stability of regional economies and communities, regional disparities in incomes, and the allocation of budgets for regional programs. These regional economic questions call for analytical systems that encompass all the forests, and all forest-related manufacturing activity within the region, and integrate them with all other sectors of economic activity to reveal the aggregate trends in such variables as employment and income and how they are affected by various actions or policies. For most purposes these models must also include the linkages among various economic sectors in order to trace the impact of a change in one sector on other sectors of the regional economy, and on the aggregate indicators of economic activity in the region.

Macroeconomic analysis. Even more broadly based models are designed to help analyse problems relating to whole economies. Macroeconomic models range from a few mathematical representations of the relationships among major components of economic activity to highly complex econometric systems, which can trace in detail the effect of changes in any one sector or policy throughout the rest of the economy.

Macroeconomic models are designed to analyse broad changes in economic activity, such as the effect of changes in interest rates, fiscal policy, and exchange rates on the general level of employment and incomes and, in the more detailed models, on particular sectors. They are often useful for investigating the impacts of events affecting the forest industry, or major changes in forest policy, but they usually involve too high a degree of aggregation to show more than the general effects felt in the economy as a whole.

Analytical models also have been designed to investigate world trade in forest products. These can reveal how shifts in international demand and supply for products can affect demand and prices for timber in particular countries and regions. Other global models are designed to examine environmental changes such as increasing carbon dioxide in the atmosphere and the "greenhouse effect," in which forests play an important role. These global systems are usually concerned primarily with ecological interrelationships and effects, but with increasing attention to measures to forestall or reverse adverse environmental changes the economic implications can be expected to receive growing attention.

RECOGNIZING IMPRECISION

Users of these analytical models at all levels face some common difficulties. These include data requirements, the specification of relationships among variables, feedback effects, and especially the simplifying assumptions that make the system complete and the problems tractable.

The models of particular concern to forest economists, at the stand and forest levels of analysis, combine the problems associated with biological systems with those of economic systems. Those seeking to analyse forestry problems often lack reliable data about the forest inventory they are concerned with, how it grows over time, and how growth can be manipulated through silviculture. Economic data about costs and prices, values of unpriced products, interest rates, and market trends over time are often meagre as well. Both types of information involve risks and uncertainties that bear on decision-making in various ways. To take advantage of the analytical techniques available, such deficiencies of information must often be bridged by rough estimates and simplifying assumptions.

As a result, the data base and empirical underpinning of even the most advanced analytical models are always more-or-less imperfect, and analysts must recognize the imprecision of their calculations. This caution is important because computers make calculations with such deceptive precision that unwarranted confidence is often put in the results. Computer-assisted analytical models are fast becoming routine and valuable tools in forest planning and decision-making, but they should be regarded primarily as means of helping to understand interrelationships in complicated systems, how specific changes or actions are transmitted through them, and how the impacts bear on the system as a whole. Their calculations can take account of more variables and in more detail than is otherwise possible. But the specific results must always be treated cautiously, in full light of the limitations of data and assumptions used to calculate them.

REVIEW QUESTIONS

1 Compare the relevant temporal and spatial scales in making decisions about (a) forest legislation, (b) a silvicultural program for a forest, and (c) fighting a forest fire.
2 Why do those who manage and use forests often lack incentives to consider the full spatial or temporal impacts of their actions? Give examples of "externalities" that can result.

3 Give examples of forestry decisions that are dealt with at the level of (a) the individual stand, (b) the forest as a whole, (c) the economic region.
4 What is an "objective function" in an analytical model?
5 How has modern computer technology advanced economic analysis of forestry decision-making? Why is it usually advisable to avoid attributing a high degree of precision to the results of computer-assisted analytical models?

FURTHER READING

Bowes, Michael D., and John V. Krutilla. 1989. *Multiple-Use Management: The Economics of Public Forestlands*. Washington, D.C.: Resources for the Future. Chapter 11

Brodie, J.D., D.M. Adams, and C. Kao. 1978. Analysis of economic impacts on thinning and rotation for Douglas-fir, using dynamic programming. *Forest Science* 24(4):513–22

Byron, R.N. 1978. Community stability and forest policy in British Columbia. *Canadian Journal of Forest Research* 8(1):61–66

Chappelle, Daniel E. 1966. Economic model building and computers in forestry research. *Journal of Forestry* 64(5):329–33. (Also in Fay Rumsey and W.A. Duerr (eds.). 1975. *Social Sciences in Forestry: A Book of Readings.* Philadelphia: W.B. Saunders. Pp. 116–22)

Conrad, Jon M., and Colin W. Clarke. 1987. *Natural Resource Economics: Notes and Problems*. Cambridge, Eng.: Cambridge University Press. 242 pp.

Cortner, H.J., and D.L. Schweitzer. 1983. Institutional limits and legal implications of quantitative models in forest planning. *Environmental Law* 13(2):493–516

Iverson, David C., and Richard M. Alston. 1986. The genesis of FORPLAN: a historical and analytical review of Forest Service planning models. USDA Forest Service General Technical Report INT-214. Intermountain Research Station, Ogden, Utah

Lundgren, Allen L. 1984. Strategies for coping with uncertainty in forest resource planning, management and use. In *New Forests for a Changing World*. Proceedings of the 1983 Convention of the Society of American Foresters. Portland, OR. Pp. 574–78

Mitchell, K.J. 1988. SYLVER: modelling the impact of silviculture on yield, lumber value and economic return. *Forestry Chronicle* 64(2): 127–31

Richardson, Harry W. 1979. *Regional Economics.* Urbana: University of Illinois Press. Chapters 8 and 9

Index